HISTORY OF MECHANISM AND MACHINE SCIENCE
Volume 3

Series Editor

MARCO CECCARELLI

Aims and Scope of the Series

This book series aims to establish a well defined forum for Monographs and Proceedings on the History of Mechanism and Machine Science (MMS). The series publishes works that give an overview of the historical developments, from the earliest times up to and including the recent past, of MMS in all its technical aspects.

This technical approach is an essential characteristic of the series. By discussing technical details and formulations and even reformulating those in terms of modern formalisms the possibility is created not only to track the historical technical developments but also to use past experiences in technical teaching and research today. In order to do so, the emphasis must be on technical aspects rather than a purely historical focus, although the latter has its place too.

Furthermore, the series will consider the republication of out-of-print older works with English translation and comments.

The book series is intended to collect technical views on historical developments of the broad field of MMS in a unique frame that can be seen in its totality as an Encyclopaedia of the History of MMS but with the additional purpose of archiving and teaching the History of MMS. Therefore the book series is intended not only for researchers of the History of Engineering but also for professionals and students who are interested in obtaining a clear perspective of the past for their future technical works. The books will be written in general by engineers but not only for engineers.

Prospective authors and editors can contact the Series Editor, Professor M. Ceccarelli, about future publications within the series at:

LARM: Laboratory of Robotics and Mechatronics
DiMSAT – University of Cassino
Via Di Biasio 43, 03043 Cassino (Fr)
Italy
E-mail: ceccarelli@unicas.it

For a list of related mechanics titles, see final pages.

Reconstruction Designs of Lost Ancient Chinese Machinery

By

Hong-Sen Yan

Department of Mechanical Engineering
National Cheng Kung University
Tainan, Taiwan

 Springer

A C.I.P. Catalogue record for this book is available from the Library of Congress.

ISBN 978-90-481-7648-9
ISBN 978-1-4020-6460-9 (e-book)

Published by Springer,
P.O. Box 17, 3300 AA Dordrecht, The Netherlands.

www.springer.com

Printed on acid-free paper

Contents

7   South-pointing Chariots

8   Walking Machines

Symbols

Index

Preface by the Series Editor, Professor M. Ceccarelli

This book is part of a book series on the History of Mechanism and Machine Science (HMMS).

This series is novel in its concept of treating historical developments with a technical approach to illustrate the evolution of matters of Mechanical Engineering that are related specifically to mechanism and machine science. Thus, books in the series will describe historical developments by mainly looking at technical details with the aim to give interpretations and insights of past achievements. The attention to technical details is used not only to track the past by giving credit to past efforts and solutions but mainly to learn from the past approaches and procedures that can still be of current interest and use both for teaching and research.

The intended re-interpretation and re-formulation of past studies on machines and mechanisms requires technical expertise more than a merely historical perspective, therefore, the books of the series can be characterized by this emphasis on technical information, although historical development will not be overlooked.

Furthermore, the series will offer the possibility of publishing translations of works not originally written in English, and of reprinting works of historical interest that have gone out of print but are currently of interest again.

I believe that the works published in this series will be of interest to a wide range of readers from professionals to students, and from historians to technical researchers. They will all obtain both satisfaction from and motivation for their work by becoming aware of the historical framework which forms the background of their research.

I would like to take this opportunity to thank the authors and editors of these volumes very much for their efforts and the time they have spent in order to share their accumulated information and understanding of the use of past techniques in the history of mechanism and machine science.

Marco Ceccarelli (Chair of the Scientific Editorial Board)
Cassino, April 2007

Preface

Ancient China was outstanding in mechanical technology before the 15th century. Numerous ingenious machines were invented. However, due to incomplete documentation and loss of finished objects, most of the original machines cannot be verified and many of the inventions did not pass down to later generations.

This book, based on the author's research and teaching experiences over the last 20 years, is devoted to presenting an innovative methodology in the area of mechanical historiography for the systematic reconstruction design of ancient Chinese machines that have been lost to time. Its purpose is to generate all possible design concepts of lost machines. If the defined and/or concluded design specifications, topological characteristics, and design requirements and constraints are feasible, one of the resulting reconstruction designs should be the original design. Such an approach provides a logical tool for historians in ancient mechanical engineering and technology to further identify the possible original designs according to proven historical archives. However, this work will not deal with the credibility of historical literary works. It supposes that the lost machines existed, and tries to demonstrate the feasibility of reconstructing the lost designs.

The book is organized in such a way that it can be used for teaching, research or self-study. Chapter 1 introduces the study, classifications, and process for the reconstruction design of ancient machinery. Chapter 2 explains the definitions of mechanisms and machines, characteristics of mechanical members and joints, concept of constrained motion, topological structure of mechanisms, and process of mechanism and machine design. Chapter 3 presents the historical development, labor-saving devices, machine elements, and mechanisms of ancient Chinese machinery. Chapter 4 provides a systematic approach for the reconstruction of all possible topological structures of lost machines. This approach utilizes the idea of creative mechanism design methodology to converge the divergent conceptions from the results of literature studies to a focused scope, and then applies the mechanical evolution and variation method to obtain feasible reconstruction designs that meet the scientific and technological standards of the subject's time period. Chapters 5–8 offer examples of lost designs, such as Zhang Heng's seismoscope, Su Song's escapement regulator, south-pointing chariots, and Lu Ban's wooden horse carriage, in a step-by-step sequence to illustrate the applications of the reconstruction design methodology provided in Chapter 4.

The book could be used as a textbook and/or supplemental reading material for courses related to history of ancient machinery and (creative) mechanism and machine design for senior and graduate students in mechanical engineering.

The author's associations with Professor Marco Cecceralli, Professor Jin-Yen Lu, and Professor Baichun Zhang during the past years have been very beneficial to the development of this book. The author wishes to thank his former graduate students, especially Dr. Tsung-Yi Lin, Dr. Chun-Wei Chen, Mr. Kuo-Hung Hsiao, Mr. Kuan-Li Lin, Mr. Cheng-Ping Chiu, Mr. Kai Hwang, Mr. Pai-Hung Chen, Mr. Huan-Wen Shen, and Ms. Chih-Chin Hung, who contributed much through their dissertations and theses. The author also wishes to thank Dr. Hsin-Te Wang, Ms. Yu-Lin Chu, and Mr. Robert Pierce, who helped him in the preparation of the book.

The reconstruction of ancient machines is the key that links ancient history and modern history of machines. With today's ever-developing technology and requirements for better and newer products, if one can reflect on the achievements of the past, one might be able to draw knowledge from the past to produce concrete objects in the present. Perhaps one might even be able to generate new ideas from understanding the past. The author believes that this book will fill the needs, both academic research and teaching, for the reconstruction design of lost ancient machinery and creative design of mechanisms and machines. Finally, comments and suggestions for improvement and revision of the book will be very much appreciated.

Hong-Sen Yan

Hong-Sen Yan 顏鴻森
Department of Mechanical Engineering
National Cheng Kung University
Tainan 70101, Taiwan
May 2007

Chapter 1 Introduction

This text explores ancient Chinese machines that have been lost to time, and proposes a methodology for systematic reconstruction to allow for their rebuilding. This chapter starts with the introduction of the study of ancient machines. It proceeds by describing the classifications of ancient Chinese machines and the process for their reconstruction design. It ends by explaining the application scope of the text.

1.1 Study of Ancient Machines

The human motivation to invent machines originated from livelihood requirements. Man-made machines have increased the production of goods, which in turn has affected lifestyle. Subsequently, social structures have upgraded due to the progressive development of machines. As a result, the invention and improvement of machines have facilitated the advancement of society.

In the long history of Chinese civilization, many ingenious machines were invented. Studies relating to the development of ancient machines have focused on people, events, objects, and causes [1–4]. Here, *people* refers to those who have contributed and affected inventions, designs, or leadership in the development of ancient machines; *events* refer to the social background, technology level, policy decision, and outcome influence in machine development; *objects* refer to the hardware, principles, and relevant literature of machines; and *causes* refer to rules, experiences, and lessons learned from the development of machines. All these four concepts are interrelated.

Studying the historical development of machines enables one to trace their paths and logics. It also leads one to understand past developmental patterns of mechanical technology. The reconstruction of ancient machines seeks to rebuild original machines by applying ancient mechanical principles, engineering, and craftsmanship. Through such an approach, the reconstructed machines can be used to demonstrate the level of mechanical technology of their times.

1.2 Classifications of Ancient Chinese Machines

Ancient China was outstanding in its mechanical technology. Numerous ingenious machines were invented before the 15th century. However, due to incomplete documentation and the loss of finished objects, most original ancient machines cannot be verified and many inventions were not passed down to later generations. For some designs, later generations could only regard these inventions as novelties, and even questioned them as being preposterous.

1.2.1 Classifications based on applications

Based on their applications, ancient Chinese machines can be classified into the following areas [5, 6]:

Labor-saving devices
Typical labor-saving devices were simple machines such as wedges, inclined planes, pulleys, levers (rods or links), wheels (pulley blocks), and their combinations.

Transmission elements
Typical transmission elements were wheels, bearings (cylinders, circular shafts), links, levers, cranks, gears, ratchets, cams, ropes, chains, screws and springs, and their combinations.

Water-drawing devices
Ancient water-drawing devices were the sharp-bottom pottery water container, shadoofs, buckets, pulley blocks, paddle blade devices, water siphons, water wheels, dragon-bone water lifts, scoop water wheels, cow-driven paddle blade machines, water-driven paddle blade machines, dragon tail machines, and others.

Agricultural devices
Ancient agricultural devices were pestle machines (tilt hammers, water-driven tilt hammers, spoon hammers, foot paddle-operated pestles, and connected pestles), grinding machines (mills, grinders, animal-driven multiple grinders, water-driven multiple grinders, and animal-driven cable grinders), crushers (stone crushers, roller crushers, and water wheelbarrows), windmills (winnowing fans), seeding machines (seed ploughs), and others.

Textile machines
Ancient textile machines were spinning machine (spinner), weaving machine, jacquard harness, and others.

Mining and metallurgy machines
Ancient mining and metallurgy machines were drills, jade polishers, the water-driven wind box, the dual-action piston wind box, and others.

Printing machines
Ancient printing machines were the rotating composing frame, rotating book shelves, and others.
Military hardware
Ancient military hardware were flamethrowers, barrel-type guns, rockets, catapults, crossbows (sight of crossbow, Zhu-ge crossbow), and others.
Aviation devices
Ancient aviation inventions were kites, airborne crafts, bamboo dragonfly, parachutes, hot air balloons, double-layered wings, and others.
Water transportation
Ancient water transportation designs were ships with paddle wheels, foot-driven paddle-wheel ships, rowing vessels, vessels that use rushing water (paddle wheel) to travel, armored ships, and others.
Land transportation
Ancient land transportation designs were war vehicles, military chariots, road carriages, man-powered twin-wheel carriages, carriages with rollers, carts with sail, single-wheel carts, south-pointing chariots, hodometers, the wooden horse carriage, the wooden cow and gliding horse, and others.
Fluid machinery
Ancient fluid machinery included revolving fans, revolving lanterns, windmills, and others.
Astronomical instruments
Ancient astronomical instruments were devices for observing the movement of the sun, solar measuring devices made from stone, sundials, compasses, telescopes, simplified armillary spheres, burning clocks, clepsydras, armillary spheres, the water-driven armillary sphere, the water-driven astronomical clock tower, the sand timer, and others.
Other devices
There were also numerous other designs, such as the sharp-bottom pottery water container, the seismograph, the lathe, the bedsheet censer, foldable umbrellas, robots, and others.

1.2.2 Classifications based on historical archives

From a restoration viewpoint, ancient Chinese machines, based on their historical archives, can be divided into three types: documented and proven, undocumented but proven, and documented but unproven [7, 8]. Here, historical archives refer to ancient manuscripts, historical artifacts, archeological data, and existing physical evidence. Ancient manuscripts refer to words and images found in official and unofficial historical records;

historical artifacts include buildings, implements, and paintings; archeological data include images and language characters on archeological findings; and existing physical evidence includes excavated ancient machines and original historical materials. Because literature on and images of artifacts only show the outer appearance but not the internal structure and dimension of parts, they are considered documented but unproven. Therefore, "documented" refers to nonphysical historical materials, while "proven" refers to the actual object.

Type I: Documented and proven

These refer to actual ancient machines with historical documentation. Generally, they are ancient machines that were widely used, some of which are excavated ancient machines with relevant literary records from their times. For instances, the wooden carriage of the Eastern Zhou Dynasty (770–256 BC) excavated from the tomb of Guo Guo (虢國) in city Sanmenxia of province Henan, the old city of Zhenghan in city Xinzheng, and the Che Ma Pit of the tomb of Zhao Qing (趙卿) in city Taiyuan of province Shanxi. Another example is the bronze arrowhead excavated from the Terra Cotta Warrior Pit of the Qin Imperial Tomb in city Xian of province Shanxsi. Relevant descriptions of these devices can be found in the book Kao Gong Ji《考工記》[9] and later annotations such as Zheng Xuan Zhu《鄭玄注》during the Eastern Han Dynasty (AD 25–219). In addition, Figure 1.1 shows the bedsheet censer (被中香爐) of Ding Huan (丁緩) in the Eastern Han Dynasty around AD 180.

Some machines such as the water wheel, water-driven tilt hammer, dragon-bone water lift, winnowing fan, and weaving machine are so practical and fully developed that they are still being used today, and relevant descriptions can often be found in historical archives. Others include the bedsheet censer and old gears.

Type II: Undocumented but Proven

These refer to excavated ancient machines that have no relevant historical documentation, such as the copper horse chariot (銅車馬) from the Qin Imperial Tomb around 210 BC and the sharp-bottom pottery water container (尖底陶瓶) excavated from the Yangshao relic site around 4,000 BC, Figure 1.2. Another unusual example is ancient Chinese paddle locks, Figure 1.3 [10, 11].

Type III: Documented but Unproven

These refer to ancient machines that have historical records but no actual evidence of existence. These machines can be further classified into those with written descriptions and illustrations, with written descriptions but without illustrations, and without written descriptions but with illustrations.

Figure 1.1 A bedsheet censer

Figure 1.2 A sharp-bottom pottery water container

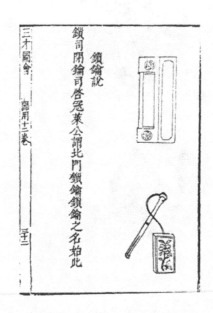

Figure 1.3 Ancient Chinese paddle locks [10, 11]

With written descriptions and illustrations
Some historical scientific literature contains written descriptions and pictorial illustrations of ancient machines. For instance, the book Wu Jing Zong Yao《武經總要》[12] during the Northern Song Dynasty (AD 960–1126) has pictorial illustrations of weapons for attacking and defending city walls, Figure 1.4. These weapons included bows, crossbows, and catapults. The book Xin Yi Xiang Fa Yao《新儀象法要》[13] also from the Northern Song Dynasty contains an illustration of an astronomical clock. The book Nong Shu《農書》[14] from the Yuan Dynasty (AD 1280–1368) has pictorial illustrations of agricultural implements such as seed ploughs and spades, as well as textile machines such as those for spinning and weaving. The book Tian Gong Kai Wu《天工開物》[15] from the Ming Dynasty (AD 1368–1644) has pictorial illustrations of daily implements and manufacturing technology such as agricultural tools, weaving machines, metallurgies, and bows and crossbows. The article Nan Chuan Ji《南船紀》[16] from the Ming Dynasty has pictorial illustrations of all kinds of civilian and military vessels such as imperial boats, military patrol boats, and bridge boats. The publication Wu Bei Zhi《武備志》[17] also from the Ming Dynasty has pictorial illustrations of weapons for attacking and defending city walls, such as rifles and cannons, as well as weapons for sea and land combat, such as war vehicles and warships.

With written descriptions but without illustrations
There are many cases with written descriptions but without illustrations. Examples are Lu Ban's (魯班) wooden horse chariot during the Eastern Zhou Dynasty, Zhang Heng's (張衡) seismograph during the Eastern Han Dynasty, Zhu Ge-liang's (諸葛亮) wooden cow and gliding horse during the period of Three Kingdoms (AD 220–280), Zhang Si-xun's (張思訓) tai ping hun yi (太平渾儀) during the Northern Song Dynasty, Yan Su's (燕肅) south-pointing chariot during the Northern Song Dynasty, and Kuo Shou-jing's (郭守敬) Deng Lou at the Da Ming Hall during the Yuan Dynasty. Most of the relevant literature focuses on the form and description of functions but have simplistic or missing narration of their structure of mechanisms. Consequently, there are problems in restoration research that require more imagination.

Without written descriptions but with illustrations
There are also historical materials with illustrations and missing written records. Examples are the picture of the ladder with wheels on the bronze pan from the late Warring Period (480–222 BC) excavated from county Ji in province Henan. Restoration of this type is similar to the case that is

documented but unproven, but is more difficult because more thorough examination of the illustrated machine is needed.

Furthermore, Type III (documented but unproven) machines can also be divided into those that have convinced most scholars, such as the seismography of Zhang Heng in the Eastern Han Dynasty around AD 132 and the astronomical mechanical clock of Su Song (蘇頌) in the North Song Dynasty in AD 1088; and those that have not convinced most scholars, such as the south-pointing chariot of the Yellow Emperor around 2500 BC and the wooden cow and gliding horse of Zhu-ge Liang in the period of Three Kingdoms around AD 230.

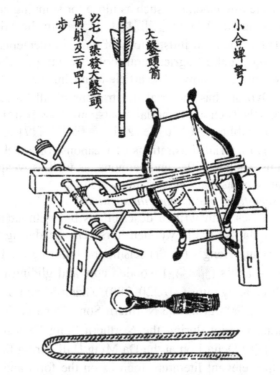

Figure 1.4 A crossbow in Wu Jing Zong Yao《武經總要》[12]

1.3 Reconstruction Design of Ancient Machines

The objective of reconstruction research is to rebuild ancient machines and to study their corresponding technology. For surviving ancient machines, the original designs are available for various studies. For lost (documented but unproven) ancient machines, since the actual machines or even pictorial illustrations are not available for verification, the reconstructed machines are often ambiguous and their correctness is difficult to prove. In general, the study of their ancient mechanical technology is even more difficult than their reconstruction.

Reconstruction research requires scientific factuality and historical objectivity in evaluating things. Only proven facts can be incorporated into the reconstruction design, and the unproven should be treated as variables. Therefore, the reconstructed design of ancient machines may not be singular, and diverse designs are normally unavoidable. Therefore, reconstructed designs can be treated as possible ancient machines belonging to the same period or as the evolutionary result of use of machines. For example, modern vehicle suspension systems are used to absorb road shocks from wheels hitting holes or bumps on the road. Figure 1.5(a)–(c) show three designs in 1980 for the rear suspension of motorcycles based on the concept of six-bar linkages by Honda (CR250R pro-link), Suzuki (RM250X full-floater), and Kawasaki (KX250 uni-trak), respectively [18]. These devices were designed for the same purpose, and the level of technology involved was similar. In fact, there can be variety in mechanical products having the same function; this was true in ancient times as well.

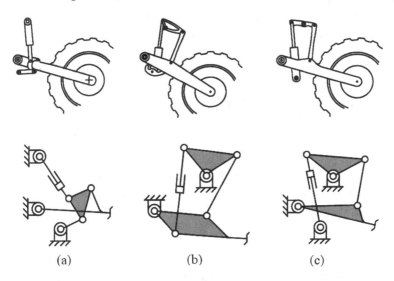

(a)   (b)   (c)

Figure 1.5 Diverse designs of motorcycle suspensions in 1980

The process for the reconstruction design of ancient machinery includes the study of historical archives, reconstruction analysis, and reconstruction synthesis, Figure 1.6 [8].

1.3.1 Study of historical archives

This step involves the study of historical archives to recognize the problem and then to identify ancient mechanical technology while using modern science and technology to define the problem for developing design specifications.

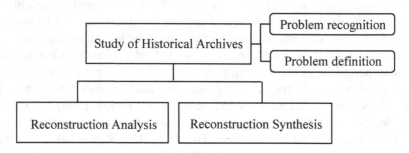

Figure 1.6 Process for reconstruction design of ancient machines

Problem recognition

Mining for information relevant to the history of ancient machines is an important foundational work, especially a thorough understanding of available archeological findings. Information to be collected should include the entire evolutionary path of the machines and existing relevant technology during that period. Attention should also be given to the names and terminologies of ancient machines. In ancient times, they may be different in different dynasties and places. For instance, the ancient Chinese water clock, named lou ke (漏刻), was referred to by different names in the past, such as qie hu (挈壺), lou (漏), tong lou (銅漏), lou hu (漏壺), ke lou (刻漏), and tong hu di lou (銅壺滴漏).

Verification, collation, and evaluation of historical archives are the first tasks toward problem recognition. Ancient literature, which accounts for the majority of historical archives, is generally simplified and may be inaccurate or exaggerated. Nevertheless, no past technological experiences and records should be ignored. For example, although there are many records on the chained-bucket mechanism in ancient literature, information on some key structures is vague, and available pictorial illustrations do not always show the structure of the chain. Since this type of machine has existed until now with very little change in its basic form, literary records and existing chained-bucket mechanisms can be used to explain its principle and structure of the design. The most credible archeological information is contained in historical artifacts. For example, drawings of sowing and weaving as well as ink paintings passed down through the ages are highly accurate reference materials. In addition, for machines that still exist today, such as water wheels and weaving machines, attention should be given to the differences between their current and ancient design and craftsmanship.

For this reason, it is necessary to refer to historical archives from various resources, as well as to correlate and verify such information in order to clarify basic issues in reconstruction research, such as principles, structures, materials, and manufacturing technology. Attention should also be given to the definition and evolution of terminologies of ancient machines. Furthermore, ancient machines in different stages and places should be collated, analyzed, and compared in order to understand their evolutionary paths.

Problem definition

Normally each past era had its own terminologies regarding engineering technology and craftsmanship. Therefore, people today might not be able to comprehend past descriptions and jargons. For this reason, it is necessary to fully understand the evolution of ancient terminologies and jargons, as well as the design, construction, technology, and craftsmanship of ancient machines. Furthermore, it is necessary to redefine ancient machines based on modern science and technology to provide a fresh and clear base for reconstruction research, especially for lost ancient machines.

After fully understanding the problem through the study of historical archives, the problem must be defined precisely so as to elicit thoughts on problem solving. Problem definition includes the development of design specifications with requirements and constraints. The purpose of specifications is to guide reconstruction design to specific domains, including special features and technical requirements such as ancient technology and craftsmanship, in order to find ways for resolving the problem.

1.3.2 Reconstruction analysis

Reconstruction design includes reconstruction synthesis and reconstruction analysis. For ancient machines that are documented and proven, as well as those that are undocumented but proven, the task of reconstruction design is mainly analysis. For lost ancient machines that are documented but unproven, the task of reconstruction design is mainly synthesis.

The purpose of reconstruction analysis is to study, test, and verify proven subjects based on modern engineering technology. For example, analytical methods such as technical testing, engineering drafting, statistical analysis, simulation verification, and scientific deductions were extensively applied to research on weapons used in ancient battles, such as bronze chariots (Figure 1.7), bows and arrows that were excavated from the burial site of the Qin Emperor [19]. Modern equipment was used to carry out accurate tests regarding the geometric parameters of the components of bronze chariots and battle weapons. The surface processing techniques used in ancient artifacts were carefully observed and analyzed to collect volumes of important raw data, which were then used for mechanical drawings. Statistical regressive analysis and numerical computing technology were used to analyze the data collected, while manufacturing simulation and experimental approaches were applied to typical components in the artifacts for comparison and verification.

Figure 1.7 A bronze chariot from the burial site of the Qin Emperor [19]

For instance, among existing artifacts is an armillary sphere from the Ming Dynasty, Figure 1.8. In 1987, the Purple Mountain Observatory and Nanjing Museum carried out reconstruction analysis on this armillary sphere [20]. Since manufacturing techniques are closely related to the metallographic structures, implements can have different metallographic structures when different manufacturing methods are used on alloys with the same composition. For this reason, 24 samples from the broken parts of the yin wei ring (陰緯環) and yang jing ring (陽經環) of the liu he yi (六合儀), the equatorial ring and tien chang ring (天常環) of the san chen yi (三辰儀), nan ji zhou zuo (南極軸座), ao yun zhu (鼇雲柱), and long zhu (龍柱) were taken for analysis. A scanning electron microscope (SEM) was used for metallographic structure verification and spectrum analysis. Based on the uniqueness of the different structures, alloy composition, and the assembled components, the performance and craftsmanship of the material of the armillary sphere were analyzed to gain an understanding of the manufacturing techniques of traditional Chinese bronze apparatuses. At the same time, measurements, tests, and verification of the structure and dimensions were performed, particularly on the shape of the components, joint techniques between the rings, and fracture design, all of which were very important in the assembly procedures. The analytical results are very helpful in the reconstruction research for various armillary spheres which were lost through the ages, such as the water-driven astronomical clock tower by Su Song in the Northern Song Dynasty.

Figure 1.8 Armillary sphere in the Purple Mountain Observatory of Nanjing

Therefore, results of reconstruction analysis are important historical information. They provide a reliable basis to identify and reestablish ancient technology and craftsmanship of a particular time period. Furthermore, they also provide invaluable reference information for the reconstruction synthesis of lost machines in the same period.

1.3.3 Reconstruction synthesis

The objective of reconstruction synthesis is to regenerate ancient machines that are consistent with historical records and the levels of ancient technology and craftsmanship subject to the developed design specifications.

Reconstruction synthesis is not new in our time but has been practiced in long-past eras with respect to even more ancient times. Most such efforts were based on verification and collation of historical manuscripts and available artifacts, and also based on a designer's knowledge, experiences, and judgments. But, very few scholars studied lost ancient machines, those with some literary records but without surviving hardware, especially based on a systematic approach.

In the past several decades, some methodologies were developed for the structural synthesis of mechanical devices. These recent engineering approaches can be divided into four types. The first is a design method based on the structure of mechanisms [21, 22]. The second is a design method based on the concept of modularization [23]. The third is a design method based on evolutionary perspective [24]. The fourth is a design method based on available database and experience [25, 26]. Each method has its advantages and disadvantages, as well as scope and domain of applications. Furthermore, they can be applied interchangeably.

Taking the reconstruction synthesis of ancient timer devices for instance, the fourth method, such as the use of Theory of Inventive Problems Solving (TRIZ), can first be used to identify possible methods for sundials, water-driven clocks, sand-driven clocks, incense time-telling, astronomical clocks, and mechanical clocks. Then the second method, which uses modular concepts, can be used on astronomical clocks to enumerate a list of system modules. Thereafter, the first method can be applied on each system module for the configuration synthesis of mechanisms. Whenever necessary, the mechanical evolution method, which is the third method, can be used to simulate design evolutions and identify all designs that are consistent with ancient mechanical principles and craftsmanship.

Taking the reconstruction synthesis of the escapement regulator of Su Song's water-driven astronomical clock tower during the Northern Song Dynasty as another example, by incorporating the creative mechanism design methodology (the first method) and the concept of mechanical evolution and variation (the third method), and based on available primitive design of the water wheel steelyard-clepsydra device, all possible design concepts that are consistent with ancient machine technology and craftsmanship are synthesized [27]. Figure 1.9 shows one of the reconstructed designs of the water wheel steelyard-clepsydra device.

In summary, the first step of reconstruction research of ancient machines is to study historical archives and apply modern engineering science and technology to redefine ancient machines and to develop design specifications. The objective is to transform the problem of reconstruction design of ancient machines into modern mechanical design, and to apply modern engineering design approaches to solve the problem. Reconstruction analysis provides important historical information for further research regarding the proven subjects. And, reconstruction synthesis generates all possible design configurations of the lost machines.

Figure 1.9 One possible reconstruction design of Su Song's water wheel steelyard-clepsydra device

1.4 Scope of the Text

The purpose of the reconstruction of ancient machines is to reproduce original machines based on principles, engineering, and craftsmanship available during that time. The reconstruction design of lost ancient machines is to regenerate possible ancient machines by applying modern design methods to ancient theories and technological level of their time period.

The book Kao Gong Ji 《考工記》 states: "A device is an accumulation of the efforts of many." 『一器於工聚焉者。』[9]. Therefore, in order to explain the actual process of historical development of machines, the models of ancient machines in various dynasties should be reconstructed. The requirements for the reconstruction of ancient machines are stringent. There is a high degree of difficulty in the reconstruction of documented but unproven ancient machines because it involves creating something from nothing, and because relevant historical archives are either incomplete, oversimplified, or exaggerated. As a result, reconstruction work for these machines focuses on reconstruction research, particularly reconstruction design. This would produce designs that are consistent with the technologies and craftsmanship existing during the times of such machines, and at the same time account for the diversity of the ancient designs.

The objective of this text is to offer an innovative approach in the field of mechanical historiography for the reconstruction design of lost machines. However, this work will not deal with the credibility of historical literary works. It supposes that the lost ancient machines existed, such as Lu Ban's wooden horse chariot during the Eastern Zhou Dynasty, and tries to present the feasibility of the lost designs.

The reconstruction of ancient machines is the key that links the ancient history and modern history of machines. With today's ever-developing technology and requirements for better and newer products, if one can reflect on the achievements of the past, one might be able to draw knowledge from the past to produce concrete models. Perhaps, one might even be able to generate new ideas from understanding the past.

References

1. Zhang, B.C., "Recollections and visions in the research of Chinese history of machines (in Chinese)," Proceedings of the 1st China-Japan International Conference on History of Machine Technology, Mechanical Industry Press, Beijing, pp. 4–7, October 1998.
   張柏春，"中國機械史研究的回顧與前瞻"，第一屆中日機械技術史國際會議論文集，機械工業出版社，北京，第4–7頁，1998年10月。
2. Chen, C.M., Lu, J.Y., and Li, J.B., "Exploring reconstruction research technology (in Chinese)," Proceedings of the 1st China-Japan International Conference on History of Machine Technology, Mechanical Industry Press, Beijing, pp. 153–159, October 1998.
   陳全明，陸敬嚴，李金伯，"復原研究技術的探索"，第一屆中日機械技術史國際會議論文集，機械工業出版社，北京，第 153–159 頁，1998年10月。
3. Lu, J.Y. and Yu, H.G., "Theoretical issues in ancient mechanical reconstruction research (in Chinese)," Proceedings of the 2nd China-Japan International Conference on History of Machine Technology, Mechanical Industry Press, Beijing, pp. 57–61, October 2000.
   陸敬嚴，虞紅根，"古代機械復原研究的幾個理論問題"，第二屆中日機械技術史國際會議論文集，機械工業出版社，北京，第 57–61 頁，2000年10月。
4. Lin, T.Y., A Systematic Reconstruction Design of Ancient Chinese Escapement Regulators (in Chinese), Ph.D. dissertation, Department of Mechanical Engineering, National Cheng Kung University, Tainan, Taiwan, December 2001.
   林聰益，古中國擒縱調速器之系統化復原設計，博士論文，國立成功大學機械工程學系，台南，台灣，2001年12月。

5. Liu, X.Z., History of Inventions in Chinese Mechanical Engineering – Vol. 1 (in Chinese), Science Press, Beijing, 1962.
劉仙洲，中國機械工程發明史(第一篇)，科學出版社，北京，1962 年。

6. Wan, D.D., Development of Chinese Mechanical Technology (in Chinese), Central Supply Agency of Cultural Subjects, Taipei, 1983.
萬迪棣，中國機械科技之發展，中央文物供應社，台北，1983 年。

7. Yan, H.S., "Technology of ancient Chinese machines and mechanisms," A tutorial at 2004 ASME International DETC & CIE Conferences, Salt Lake City, Utah, 6 October 2004.

8. Lin, C.Y. and Yan, H.S., "Approach and procedure for reconstruction research for ancient machinery (in Chinese)," Journal of Guangxi University of Nationalities, Science edition, Nanning, Guangxi, Vol. 12, No. 2, pp. 37–42, May 2006.
林聰益，顏鴻森，"古機械復原研究的方法與程序"，廣西民族學院學報 (自然科學版)，南寧，廣西，第 12 卷，第 2 期，第 37–42 頁，2006 年 05 月。

9. Kao Gong Ji (in Chinese), annotated by Zheng Xuan (Han Dynasty), commentaries by Jia Gong-yan (Tang Dynasty), collated by Ruan Yuan (Qin Dynasty), Notes and Commentaries on Zhou Li, Chapter 41, Da Hua Publishing House, Taipei, 1989.
《考工記》：鄭玄[漢朝]注，賈公彥[唐朝]疏，阮元[清朝]校勘，周禮注疏，卷四十一，大化出版社，台北，1989 年。

10. San Cai Tu Hui (in Chinese) by Wang Qi (Ming Dynasty), Zhuang Yan Culture Co., Tainan, Taiwan, 1995.
《三才圖會》：王圻[明朝]撰，莊嚴文化事業公司，台南，台灣，1995 年。

11. Yan, H.S., The Beauty of Ancient Chinese Locks, Ancient Chinese Machines Cultural Foundation, Tainan, Taiwan, May 2003.

12. Wu Jing Zong Yao (in Chinese) by Zeng Gong-liang (Northern Song Dynasty), The Commercial Press, Shanghai, 1935.
《武經總要》：曾公亮[北宋]撰，商務印書館，上海，1935 年。

13. Xin Yi Xiang Fa Yao (in Chinese) by Su Song (Northern Song Dynasty), Taiwan Commercial Press, Taipei, 1969.
《新儀象法要》：蘇頌[北宋]撰，台灣商務印書館，台北，1969 年。

14. Nong Shu (in Chinese) by Wang Zhen (Yuan Dynasty), Taiwan Commercial Press, Taipei, 1968.
《農書》：王禎[元朝]撰，台灣商務印書館，台北，1968 年。

15. Tian Gong Kai Wu (in Chinese) by Song Ying-xing (Ming Dynasty), Taiwan Commercial Press, Taipei, 1983.
《天工開物》：宋應星[明朝]撰，天工開物，台灣商務印書館，台北，1983 年。

16. Nan Chuan Ji (in Chinese) by Shen Zi-you (Ming Dynasty), General Collection of Chinese Classics of Science and Technology, Volume on Technology, Chapter 1, Henan Education Publishing House, Zhangzhou, 1993.

《南船紀》；沈子由[明朝]撰，中國科學技術典籍通彙, 技術卷一，河南教育出版社，鄭州，1993 年。

17. Wei's Records of the Spring and Autumn Period (in Chinese) by Sun Sheng (Eastern Jin Dynasty), Editorial Committee on the Collection of Chinese History, Library, Si Chuan University, Chengdu, 1993.
《魏氏春秋》；孫盛[東晉]撰，中國史集成編委會，四川大學圖書館，成都，1993 年。

18. Yan, H.S., Creative Design of Mechanical Devices, Springer, Singapore, October 1998.

19. Yang, C., "Research and verification of mechanical engineering in the Qin Dynasty (in Chinese)," Journal of the Northwest Agriculture University, Special edition, Xianyang, Vol. 23, p. 23, 1995.
楊青，秦代機械工程的研究與考證專輯，西北農業大學學報，咸陽，第 23 卷，第 23 頁，1995 年。

20. Wu, K.I., Wang, C.C., and Li, H.H., "Research on the construction techniques of the armillary sphere and the simplified armillary sphere (in Chinese)," Southeast Culture, Nanjing, Vol. 6, pp. 97–111, 1994.
吳坤儀，王金潮，李秀會，渾儀、簡儀製作技術的研究，東南文化，南京，第 6 期，第 97–111 頁，1994 年。

21. Freudenstein, F. and Maki, F., "The creation of mechanism according to kinematic structure and function," Environment and Planning B, Vol. 6, pp. 375–391, 1979.

22. Yan, H.S., "A Methodology for creative mechanism design," Mechanism and Machine Theory, Vol. 27, No. 3, pp. 235–242, 1992.

23. Kota, S. and Chiou, S.J., "Conceptual design of mechanisms based on computational synthesis and simulation of kinematic building blocks," Research in Engineering Design, Vol. 4, pp. 75–87, 1992.

24. Liang, Z.J. and Liang, S., 2000, "A new angle of view in machinery history studies – drawing up evolution pedigree and innovation," Proceedings of HMM2000 International Symposium on History of Machines and Mechanisms, Cassino, Italy, Kluwer Academic Publishers, Dordrecht, The Netherlands, pp. 283–290, 2000.

25. Altshuller, G.S., Creativity as an Exact Science, Gordon & Breach, New York, 1998.

26. Terninko, J., Zusman, A., and Zlotin, B., Systematic Innovation: An Introduction to TRIZ (Theory of Inventive Problem Solving), St. Lucie Press, New York, 1998.

27. Yan, H.S. and Lin, T.Y., "A systematic approach to the reconstruction of ancient Chinese escapement regulators," Proceedings of ASME 2002 Design Engineering Technical Conferences – the 27th Biennial Mechanisms and Robotics Conferences, Montreal, 29 September–2 October 2002.

Chapter 2 Mechanism and Machine

One of the fundamental concerns in designing a machine is its mechanism for achieving a desired motion. The task of reconstruction design of lost ancient machines is mainly the design of mechanisms. This chapter presents the definitions of mechanisms and machines, the characteristics of mechanical members and joints, the definitions of kinematic link chains, the concept of constrained motion, the identification of the topological structure of mechanisms, and the process of mechanism and machine design [1, 2].

2.1 Definitions

A mechanism is an assembly of mechanical members connected by joints, and these members are so formed and connected such that they transmit constrained motions by moving upon each other. Four-bar linkages used in various applications for transmitting motions are typical examples of mechanisms. Figure 2.1(a) shows a four-bar linkage for guiding an automobile hood, and Figure 2.1(b) shows a four-bar linkage, named in ancient China the jie chi, for drawing parallel lines.

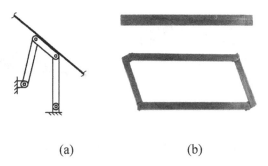

(a)   (b)

Figure 2.1 Applications of four-bar linkages

A machine is a piece of equipment, mechanical in nature, designed for producing an effective work output or for conserving mechanical energy. In general, it consists of one or more mechanisms, has a certain type of power input, and has an adequate control system in order to serve a special purpose or perform a special function. Typical machines are working machines which convert mechanical energy to effective work outputs, such as machine tools, forklifts, generators, and compressors. Figure 2.2 shows a modern vertical machining center and its automatic tool changer. Prime movers like internal combustion engines, steam engines, turbines, and electrical motors are also machines which transform other forms of energy (such as wind, heat, water, electricity, and so on) to mechanical energy as the driving power of working machines. Figure 2.3 shows a jet-engine turbine for a commercial aircraft. Machines should have suitable controlling devices, such as human power control, hydraulic control, pneumatic control, electrical control, electronic control, and computer control for effectively producing the required motion and work. Figure 2.4 shows a hand-operated paddle machine used in ancient China [3]. It is a machine with human power as the input, and wooden paddle chains and sprockets as the mechanism.

A structure is an assembly of mechanical members connected by joints, and these members are so formed and connected that they transmit forces without any relative motion. A bridge is a structure. Aircraft-landing gears, for the purpose of absorbing impact forces during landing, are also structures; however, they become mechanisms during the period between gear up and gear down.

Figure 2.2 A machining center and its automatic tool changer

Figure 2.3 A jet-engine turbine

Figure 2.4 A hand-operated paddle blade machine [3]

Every mechanism and machine has a structure member known as the frame or ground link for guiding the motions of some mechanical members, for transmitting forces or for withstanding stresses. A frame can be a piece of a mechanical member or an overconstrained assembly of several mechanical members. For instance, the frame of a machine tool is the ground member and it is a structure.

2.2 Mechanical Members

Mechanical members are resistant bodies that collectively form mechanisms and machines. They can be rigid members, flexible members, or compression members. Compression members (such as airs or fluids) and those for the purpose of fastening two or more members together (such as shafts, keys, and rivets) that play no role in the reconstruction design of ancient machines are not of interest here. Only those members whose function is to provide possible relative motion with others are presented.

There are numerous types of mechanical members. The following are functional descriptions of basic mechanical members of machines.

Link

A link (K_L) is a rigid member for holding joints apart and for transmitting motions and forces. Generally, any rigid mechanical member is a link. Links can be classified based on the number of incident joints. A separated link is one with zero incident joints. A singular link is one with one incident joint. A binary link is one with two incident joints. A ternary link is one with three incident joints. A quaternary link is one with four incident links. An L_i-link is one with i incident joints. Graphically, a link with i incident joints is symbolized by a shaded, i-sided polygon with small circles on the vertices indicating incident joints. Furthermore, two binary links in a series is called a dyad.

Slider

A slider (K_P) is a link that has either rectilinear or curvilinear translation. Its purpose is to provide a sliding contact with adjacent member.

Roller

A roller (K_O) is a link for the purpose of providing rolling contact with an adjacent member. A wheel is basically a roller.

Cam

A cam (K_A) is an irregularly shaped link that serves as a driving member and it imparts a prescribed motion to a driven link called follower (K_{Af}). Cams can be classified as wedge cams, disk cams, cylindrical cams, barrel cams, conical cams, spherical cams, roller gear cams, and others.

Gear
Gears (K_G) are links that are used, by means of successively engaging teeth, to provide positive motion from a rotating shaft to another that rotates, or from a rotating shaft to a body that translates. Gears can be classified as pin gears, spur gears, bevel gears, helical gears, and worm and worm gears.

Screw
Screws (K_H) are used for transmitting motions in a smooth and uniform manner. They may also be thought of as linear actuators that transform rotary motion into linear motion.

Belt
Belts (K_B) are tension members for power transmissions and conveyers. They obtain their flexibility from material distortion, and motion is usually transmitted by means of friction between the belts and their corresponding pulleys (K_U). Belts can be classified as flat belts, V belts, and timing belts.

Chain
Chains (K_C) are also tension members for power transmissions and conveyors. They are made from small rigid parts that are joined in such a manner as to permit relative motion of the parts, and motion is usually transmitted by positive means, such as sprockets (K_K). Chains can be classified as hosting chains, conveying chains, and power transmission chains.

Actuator
An actuator (K_T) consists of a piston (K_I) and a cylinder (K_Y) with a kind of compression member bounded by the piston and the cylinder. Its purpose is to provide a damping action between the members adjacent to the actuator. Shock absorbers of vehicles are typical actuators.

Spring
Springs (K_S) are flexible members. They are used for storing energy, applying forces, and making resilient connections. Springs can be classified as wire springs, flat springs, and special-shaped springs.

2.3 Joints

In order for mechanical members to be useful, they must be connected by certain means. That part of a mechanical member that is connected to a part of another member is called a pairing element. Two elements that belong to two different members and are connected together form a kinematic pair or joint.

Kinematic pairs or joints are categorized according to the degrees of freedom, the type of motion, the type of contact, and the type of joints. These features are introduced as follows:

Degrees of freedom

The number of degrees of freedom is the number of independent parameters needed to specify the relative positions of the pairing elements of a joint. An unconstrained pairing element has six degrees of freedom including three translational and three rotational degrees of freedom. When a pairing element connects to another pairing element and forms a joint, a constraint is imposed and the motion of the original member is reduced by one or more degrees of freedom. Hence, a joint has a maximum of five degrees of freedom and a minimum of one degree of freedom. The topic on degrees of freedom and constrained motion will be further discussed in Section 2.5.

Type of motion

Type of motion refers to the motion of a point on a pairing element relative to another pairing element of a joint. The motion can be rectilinear or curvilinear, planar or curved, or spatial.

Type of contact

Point contacts, line contacts, and surface contacts are types of contact between two pairing elements.

Type of joints

In what follows, the functional descriptions of basic joints are introduced and the corresponding schematic representations of joints are shown in corresponding figures.

Revolute joint

For a revolute joint (J_R), Figure 2.5(a), the relative motion between two incident members is rotation about an axis. It has one degree of freedom with circular motion and surface contact.

Prismatic joint

For a prismatic joint (J_P), Figure 2.5(b), the relative motion between two incident members is translation along an axis. It has one degree of freedom with rectilinear motion and surface contact.

Rolling joint

For a rolling joint (J_O), Figure 2.5(c), the relative motion between two incident members is pure rolling without slipping. It has one degree of freedom with cycloid motion and line contact.

Cam joint

For a cam joint (J_A), Figure 2.5(d), the relative motion between two incident members is the combination of rolling and sliding. It has two degrees of freedom with curvilinear motion and line contact.

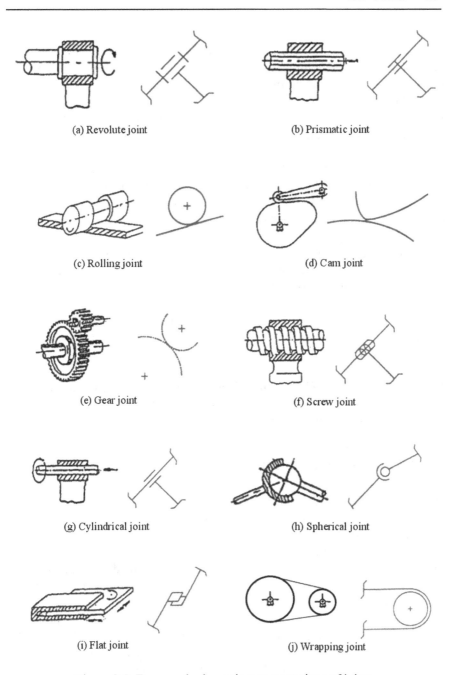

(a) Revolute joint

(b) Prismatic joint

(c) Rolling joint

(d) Cam joint

(e) Gear joint

(f) Screw joint

(g) Cylindrical joint

(h) Spherical joint

(i) Flat joint

(j) Wrapping joint

Figure 2.5 Types and schematic representations of joints

Gear joint

For a gear joint (J_G), Figure 2.5(e), the relative motion between two incident members is the combination of rolling and sliding. It has two degrees of freedom with curvilinear motion and line contact.

Screw joint

For a screw joint (J_H), Figure 2.5(f), the relative motion between two incident members is helical.

Cylindrical joint

For a cylindrical joint (J_C), Figure 2.5(g), the relative motion between two incident members is the combination of a rotation about an axis and a translation parallel to the same axis. It has two degrees of freedom with curvilinear motion and surface contact.

Spherical joint

For a spherical joint (J_S), Figure 2.5(h), the relative motion between two incident members is spherical. It has three degrees of freedom with spherical motion and surface contact.

Flat joint

For a flat joint (J_F), Figure 2.5(i), the relative motion between two incident members is planar. It has three degrees of freedom with planar motion and surface contact.

Wrapping joint

For a wrapping joint (J_W), Figure 2.5(j), there is no relative motion between two incident members. However, one of the members (pulley or sprocket) rotates about its center.

2.4 Mechanisms and (Link) Chains

When several links are connected together by joints, they are said to form a link chain, or just a chain in short. An (N_L, N_J) chain refers to a chain with N_L links and N_J joints.

A walk of a chain is an alternating sequence of links and joints beginning and ending with links, in which each joint is incident with the two links immediately preceding and following it. For example, for the (5, 4) chain shown in Figure 2.6(a), link 1 – joint b – link 4 – joint d – link 3 – joint d – link 4 is a walk. A path of a chain is a walk in which all the links are distinct. For example, for the (5, 4) chain shown in Figure 2.6(a), link 1 – joint b – link 4 – joint d – link 3 is a path. If any two links of a chain can be joined by a path, the chain is said to be connected; otherwise the chain is disconnected. Figure 2.6(a) shows a (5, 4) disconnected chain with a separated link (link 5), and Figure 2.6(b) shows a (5, 5) connected chain with a singular link (link 5). If every link in the chain is connected to at least two other links, the chain forms one or several closed loops and is called a closed chain. A connected chain that is not closed is an open chain. A bridge-link in a chain is a link whose removal results in a disconnected chain. Figure 2.6(c) shows a (6, 7) closed chain with a bridge-link (link 4). The connected chain shown in Figure 2.6(b) is also an open chain.

A kinematic chain generally refers to a movable chain that is connected, closed, without any bridge-link, and with revolute joints only. If one of the links in a kinematic chain is fixed as the ground link (K_F), it is a mechanism. Figure 2.6(d) shows a (6, 7) kinematic chain, and Figure 2.6(e) shows its corresponding mechanism obtained by grounding link 1 in the

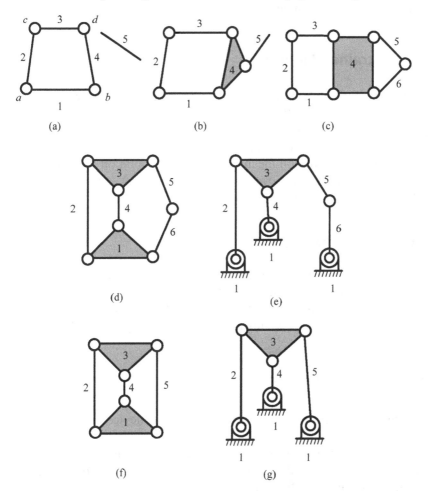

Figure 2.6 Type of (link) chains, mechanisms, and structures

chain. A generalized kinematic chain consists of generalized links connected by generalized joints, i.e., the types of links and joints are not specified. For example, if the types of links and joints for the (6, 7) kinematic chain shown in Figure 2.6(d) are not specified, it becomes a (6, 7) generalized kinematic chain. The concept of generalized kinematic chains will be presented in detail in Section 4.4.

A rigid chain refers to an immovable chain that is connected, closed, without any bridge-link, and with revolute joints only. If one of the links in a rigid chain is fixed or grounded, it is a structure. Figure 2.6(f) shows a (5, 6) rigid chain, and Figure 2.6(g) shows its corresponding structure by grounding link 1 in the chain.

2.5 Constrained Motion

The number of degrees of freedom (F) of a mechanism determines how many independent inputs the design must have in order to fulfill a useful engineering purpose. A mechanism with a positive number of degrees of freedom and with the same number of independent inputs has constrained motion. Constrained motion means that when any point on an input member of the mechanism is moved in a prescribed way, all other moving points of the mechanism have uniquely determined motions.

2.5.1 Planar mechanisms

For planar mechanisms, a member has three degrees of freedom consisting of translational motions along two mutually perpendicular axes and a rotational motion about any point. The number of degrees of freedom, F_p, of a planar mechanism with N_L members and N_{Ji} joints of type i is:

$$F_p = 3(N_L - 1) - \sum N_{Ji} C_{pi} \tag{2.1}$$

where C_{pi} is the number of degrees of constraint of i-type joint.

[Example 2.1]
Calculate the number of degrees of freedom for the device jie chi shown in Figure 2.1.

This is a planar mechanism with four members and four revolute joints. Therefore, $N_L = 4$, $C_{pR} = 2$, $N_{JR} = 4$, and $C_{pP} = 2$. Based on Equation (2.1), the number of degrees of freedom, F_p, of this mechanism is:

$$F_p = 3(N_L - 1) - (N_{JR}C_{pR})$$
$$= (3)(4-1) - (4)(2)$$
$$= 9 - 8$$
$$= 1$$

Therefore, the motion of this mechanism is constrained.

[Example 2.2]
Calculate the number of degrees of freedom for the (5, 6) planar mechanism shown in Figure 2.7. The actuator (members 4 and 5) is the input and the follower (member 2) is the output link.

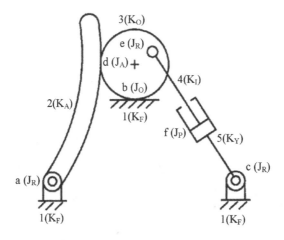

Figure 2.7 A (5, 6) planar mechanism

This is a planar mechanism with five members (ground link K_F, link 1; follower K_{Af}, link 2; roller K_O, link 3; piston K_I, link 4; and cylinder K_Y, link 5) and six joints consisting of three revolute joints (J_R; a, c, and e), one prismatic joint (J_P; f), one rolling joint (J_O; b), and one cam joint (J_A; d). Therefore, $N_L = 5$, $C_{pR} = 2$, $N_{JR} = 3$, $C_{pP} = 2$, $N_{JP} = 1$, $C_{pO} = 2$, $N_{JO} = 1$, $C_{pA} = 1$, and $N_{JA} = 1$. Based on Equation (2.1), the number of degrees of freedom, F_p, of this mechanism is:

$$F_p = 3(N_L - 1) - (N_{JR}C_{pR} + N_{JP}C_{pP} + N_{JO}C_{pO} + N_{JA}C_{pA})$$
$$= (3)(5 - 1) - [(3)(2) + (1)(2) + (1)(2) + (1)(1)]$$
$$= 12 - 11$$
$$= 1$$

Therefore, the motion of this mechanism is constrained.

[Example 2.3]
Calculate the number of degrees of freedom for the horizontal tail control
mechanism with two independent inputs of an aircraft shown in Figure 2.8
in which member 2 is an input (I) from the control stick, the actuator
(members 8 and 9) is another input (II) for the purpose of stability aug-
mentation, and member 7 is the output link to the horizontal tail.

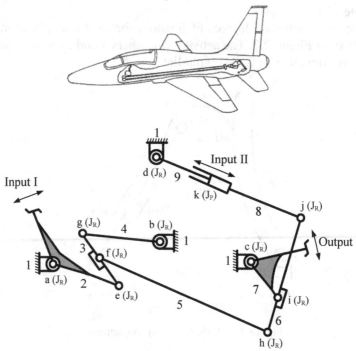

Figure 2.8 An aircraft horizontal tail control mechanism

This is a planar mechanism with nine members (links 1–9) and eleven
joints consisting of ten revolute joints (a–j) and one prismatic joint (k).
Therefore, $N_L = 9$, $C_{pR} = 2$, $N_{JR} = 10$, $C_{pP} = 2$, and $N_{JP} = 1$. Based on Equation (2.1), the number of degrees of freedom, F_p, of this mechanism is:

$$F_p = 3(N_L - 1) - (N_{JR}C_{pR} + N_{JP}C_{pP})$$
$$= (3)(9 - 1) - [(10)(2) + (1)(2)]$$
$$= 24 - 22$$
$$= 2$$

Therefore, the motion of this mechanism is constrained.
    If the stability augmented system, that is, input II, is not activated,
links 8 and 9 have no relative motion to each other. It then becomes a

mechanism with eight members and ten revolute joints, and its number of degrees of freedom, F_p, is:

$$F_p = 3(N_L - 1) - N_{JR}C_{pR}$$
$$= (3)(8-1) - (10)(2)$$
$$= 21 - 20$$
$$= 1$$

Therefore, the motion of this mechanism is still constrained.

2.5.2 Spatial mechanisms

For spatial mechanisms, a member has six degrees of freedom consisting of translational motions along three mutually perpendicular axes and three rotational motions about these axes. The number of degrees of freedom, F_s, of a spatial mechanism with N_L members and N_{Ji} joints of type i is:

$$F_s = 6(N_L - 1) - \sum N_{Ji}C_{si} \qquad (2.2)$$

where C_{si} is the number of degrees of constraint of i-type joint.

[Example 2.4]
Explain if the McPherson strut suspension mechanism (Figure 2.9) commonly used in automobiles has constrained motion. The input of this device is from the wheel to member 3.

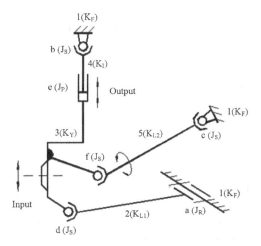

Figure 2.9 McPherson strut suspension mechanism

This is a spatial mechanism with five members (frame K_F, link 1; connecting link K_{L1}, link 2; wheel link and cylinder K_Y, link 3; piston K_I, link 4; connecting link K_{L2}, link 5) and six joints consisting of one revolute joint (a), one prismatic joint (e), and four spherical joints (b, c, d, and f). Therefore, $N_L = 5$, $C_{sR} = 5$, $N_{JR} = 1$, $C_{sP} = 5$, $N_{JP} = 1$, $C_{sS} = 3$, and $N_{JS} = 4$. Based on Equation (2.2), the number of degrees of freedom, F_s, of this mechanism is:

$$
\begin{aligned}
F_s &= 6(N_L - 1) - (N_{JR}C_{sR} + N_{JP}C_{sP} + N_{JS}C_{sS}) \\
&= (6)(5 - 1) - [(1)(5) + (1)(5) + (4)(3)] \\
&= 24 - 22 \\
&= 2
\end{aligned}
$$

This is still a useful device, since the rotation of member 5 about the axis through the centers of spherical joints c and f is an extra degree of freedom that does not affect the input–output relation of the system.

[Example 2.5]
Calculate the number of degrees of freedom for the ancient Chinese mill for removing rice hulls shown in Figure 2.10.

Since the two ropes are designed for the purpose of providing an efficient input through human power and are symmetrical, this device can be analyzed as a spatial mechanism with four members (the ground link K_F, member 1; the rope K_R, member 2; the horizontal bar and connecting rod K_{L1}, member 3; and the crank and the grinding stone K_{L2}, member 4). There are four joints consisting of two spherical joints (J_S; joint a and joint b) and two revolute joints (J_R; joint c and joint d). Therefore, $N_L = 4$, $C_{sR} = 5$, $N_{JR} = 2$, $C_{sS} = 3$, and $N_{JS} = 2$. Based on Equation (2.2), the number of degrees of freedom, F_s, of this mechanism is:

$$
\begin{aligned}
Fs &= 6(N_L - 1) - (N_{JR}C_{sR} + N_{JS}C_{sS}) \\
&= (6)(4 - 1) - [(2)(5) + (2)(3)] \\
&= 18 - 16 \\
&= 2
\end{aligned}
$$

2.6 Topological Structures

Two mechanisms are said to be isomorphic if they have the same topological structures. The topological structure of a mechanism is characterized by its types and numbers of links and joints, and the incidences between them. The isomorphism of mechanisms can be identified based on the concept of a matrix named topology matrix, which is a powerful tool for the representation of the topological structures of various chains.

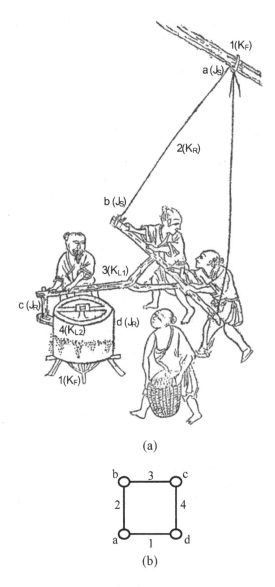

(a)

(b)

Figure 2.10 An ancient Chinese mill [4]

The topology matrix, M_T, of an (N_L, N_J) chain or mechanism is an N_L by N_L matrix. Its diagonal element is $e_{ii} = u$ if the type of member i is u. Its upper off-diagonal entry is $e_{ik} = v$ $(i < k)$ if the type of the joint incident to members i and k is v, and its lower off-diagonal entry is $e_{ki} = w$ if the assigned name of the joint is w; and $e_{ik} = e_{ki} = 0$ if members i and k are not adjacent.

[Example 2.6]
Identify the topological structure of the $(5, 6)$ planar mechanism shown in Figure 2.7.

This mechanism has five members and six joints. K_F (member 1) is the ground link, K_{Af} (member 2) is the output link, K_O (member 3) is a roller, K_I (member 4) is a piston, and K_Y (member 5) is a cylinder. The joint (a) incident to K_F and K_{Af} is a revolute joint (J_R); the joint (b) incident to K_F and K_O is a rolling joint (J_O); the joint (c) incident to K_F and K_Y is a revolute joint (J_R); the joint (d) incident to K_{Af} and K_O is a cam joint (J_A); the joint (e) incident to K_O and K_I is a revolute joint (J_R); and the joint (f) incident to K_I and K_Y is a prismatic joint (J_P).

The topology matrix, M_T, of this mechanism is:

$$M_T = \begin{bmatrix} K_F & J_R & J_O & 0 & J_R \\ a & K_{Af} & J_A & 0 & 0 \\ b & d & K_O & J_R & 0 \\ 0 & 0 & e & K_I & J_P \\ c & 0 & 0 & f & K_Y \end{bmatrix}$$

[Example 2.7]
Identify the topological structure of the McPherson strut suspension mechanism shown in Figure 2.9.

This mechanism also has five members and six joints. Member 1 (K_F) is the ground, member 2 (K_{L1}) is a connecting link, member 3 (K_Y) is the wheel link and the cylinder of the shock absorber, member 4 (K_I) is the piston of the shock absorber, and member 5 (K_{L2}) is another connecting link. The joint (a) incident to K_F and K_{L1} is a revolute joint (J_R); the joint (b) incident to K_F and K_I, and the joint (c) incident to K_F and K_{L2} are spherical joints (J_S); the joint (d) incident to K_{L1} and K_Y is also a spherical joint (J_S); the joint (e) incident to K_Y and K_I is a prismatic joint (J_P); and the joint (f) incident to K_Y and K_{L2} is another spherical joint (J_S).

The topology matrix, M_T, of this mechanism is:

$$M_T = \begin{bmatrix} K_F & J_R & 0 & J_S & J_S \\ a & K_{L1} & J_S & 0 & 0 \\ 0 & d & K_Y & J_P & J_S \\ b & 0 & e & K_I & 0 \\ b & 0 & f & 0 & K_{L2} \end{bmatrix}$$

[Example 2.8]

Identify the topological structure of the ancient Chinese mill shown in Figure 2.10.

This mechanism has four members and four joints. Member 1 (K_F) is the ground, member 2 (K_R) is the rope, member 3 (K_{L1}) is the connecting link, and member 4 (K_{L2}) is the crank. The joint (a) incident to K_F and K_R is a spherical joint (J_S); the joint (b) incident to K_R and K_{L1} is also a spherical joint (J_S); the joint (c) incident to K_{L1} and K_{L2} is a revolute joint (J_R); and the joint (d) incident to K_{L2} and K_F is also a revolute joint (J_R).

The topology matrix, M_T, of this mechanism is:

$$M_T = \begin{bmatrix} K_F & J_S & 0 & J_R \\ a & K_R & J_S & 0 \\ 0 & b & K_{L1} & J_R \\ d & 0 & c & K_{L2} \end{bmatrix}$$

2.7 Mechanism and Machine Design

The process for designing mechanisms and machines can be divided into the following seven steps as outlined in Figure 2.11.

Step 1. Problem definition

The first step in the design process is to define the problem, which includes listing all the specifications for the mechanism and the machine to be designed. The specifications are the input and output quantities such as the type of driving power, the required motion of the output member, the strength and rigidity of the members, and the amount of effective work produced. These should be systematically written down by designers to further enhance the functional design of the machine.

Step 2. Structural synthesis

After the output function and input type of a machine have been defined, the next step is to synthesize feasible topological structures based on the characteristics of the mechanism, and the design requirements and design constraints for achieving a constrained motion.

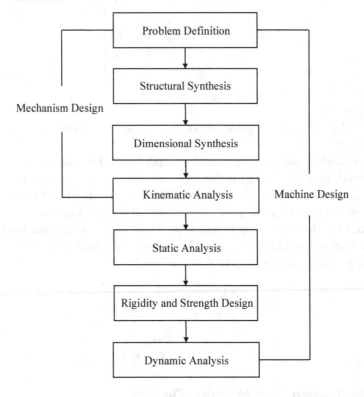

Figure 2.11 Process of mechanism and machine design

Step 3.   Dimensional synthesis

The purpose of this step is to synthesize the geometric dimensions of members between joints such that the input state of motion (position, velocity and acceleration) corresponds to the required output state of motion. The key issue of this step is to determine the geometric relation and the relative motion of the machine members without considering the size, the mass, and the loading condition.

Step 4.   Kinematic analysis

This step validates the synthesized mechanisms from Step 3 by determining if the input and output motion relations satisfy the requirements. These include the angular and the linear position, velocity and acceleration of the machine members and points of interest in the mechanisms. And, they are used for dynamic analysis.

Step 5. Static analysis

Based on the theory of statics and known static loading, this step is to analyze the static forces acting on every joint of the machine members in all possible positions. It also studies the required force or torque of the input members.

Step 6. Rigidity and strength design

Based on the theory of the strength of materials, known static loading and selection of materials, this step is to determine the size of machine members for ensuring that the members are sufficiently strong and rigid to safely withstand the imposed loading.

Step 7. Dynamic analysis

This step studies dynamic problems that arise with the motion and the mass of machine members. These problems include dynamic loading, inertia force, shaking force, shaking moment, dynamic balance, and dynamic response, etc. In this step, the strength and rigidity of the designed machine members from the previous steps are rechecked to ensure that they are within safety limits.

Mechanism design concerns only the first four steps, and machine design includes all seven steps. No step is independent since in the process of design, if any of the steps does not satisfy specifications, the results from the previous steps would have to be suitably revised. Other than the above-stated seven steps, the complete machine design process includes selection of a driving power, design of a control system, industrial design, the selection of material and treatments, thermal and fluid effect consideration, assembly and testing, and so on. The reconstruction design of lost ancient machines focuses on Steps 1 and 2, i.e., the problem definition and structure synthesis of mechanisms.

2.8 Structural Synthesis of Mechanisms

The reconstruction design of lost ancient machines is basically to synthesize all possible topological structures of mechanisms. Structural synthesis is the process of synthesizing the topological structure of mechanisms including the number of degrees of freedom, the numbers and types of machine members and joints, and the atlas of (generalized) kinematic chains. The process for the structural synthesis of mechanisms can be divided into the following steps as outlined in Figure 2.12.

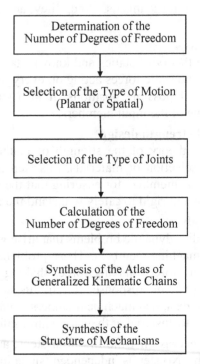

Figure 2.12 Process of structural synthesis of mechanisms

Step 1. Determination of the number of degrees of freedom

The number of inputs is a known condition and is decided by the purpose of the task. If no special requirement is specified, the number of degrees of freedom is taken as the number of independent inputs of the mechanism.

Step 2. Selection of the type of motion

To determine whether the mechanism is planar or spatial, various factors such as specifications, design requirements, design constraints and the designer's own judgments (e.g., the positions and the type of motion of the input and output members) have to be taken into consideration. If a planar mechanism is chosen, Equation (2.1) should be applied for the structural synthesis. Conversely, Equation (2.2) is applied for a spatial mechanism.

Step 3. Selection of the type of joints

This is based on the objective of the task, the type of motion, and the designers' judgments. Generally, a joint with one degree of freedom is chosen if no special requirement is specified.

Step 4. Calculation of the numbers of members and joints

Once Steps 2 and 3 are accomplished, Equation (2.1) or Equation (2.2) is employed to calculate the required numbers of members and joints.

Step 5. Synthesis of the atlas of generalized kinematic chains

The synthesis of the atlas of (N_L, N_J) generalized kinematic chains with N_L links and N_J joints is beyond the scope of this book. Figures 4.7–4.17 provide commonly used atlases of generalized kinematic chains.

Step 6. Synthesis of the structure of mechanisms

The last step is to synthesize the structure of mechanisms by assigning specific types of members and joints in every (N_L, N_J) (generalized) kinematic chain obtained in Step 5 subject to design requirements and constraints.

[Example 2.9]

Carry out the structural synthesis of a planar mechanism with one independent input and calculate the number of required links and joints.

1. Since the mechanism has one independent input with no other specifications, the number of degree of freedom is taken as one.
2. Since the mechanism is specified as planar, based on Equation (2.1),

$$F_p = 3 (N_L - 1) - \Sigma N_{Ji} C_{pi} = 1 \tag{2.3}$$

3. Since joints in a planar mechanism with one degree of freedom are revolute joints ($C_{pR} = 2$), prismatic joints ($C_{pP} = 2$) or rolling joints ($C_{pO} = 2$), based on Equations (2.1) and (2.3),

$$N_{JR} + N_{JP} + N_{JO} = (3N_L - 4)/2 \tag{2.4}$$

4. Solving Equation (2.4) for the case of the smallest number of links, i.e., $N = 4$, results in the following solutions:

N_{JR}	0	0	0	0	0	1	1	1	1	2	2	2	3	3	4
N_{JP}	0	1	2	3	4	0	1	2	3	0	1	2	0	1	0
N_{JO}	4	3	2	1	0	3	2	1	0	2	1	0	1	0	0

Therefore, the number of joints is $N_J = N_{JR} + N_{JP} + N_{JO} = 4$.

[Example 2.10]

Carry out the structural synthesis for motorcycle rear suspension mechanisms with planar six-bar linkages and a shock absorber.

1. The only input to the rear suspension mechanism of a motorcycle is from the motion along the ground. As such, the number of independent input is one and the number of degree of freedom is also one.
2. It is a planar mechanism.
3. To simplify the design and construction of the mechanism, the joints are chosen as revolute joints. Since a suspension mechanism must have a prismatic joint for the shock absorber, based on Equation (2.1), the number of revolute joints, N_{JR}, is:

$$N_{JR} = (3N_L - 4)/2 \qquad (2.5)$$

4. If the number of members does not exceed eight, based on Equation (2.5), the following solutions can be obtained,

N_L	4	6	8
N_J	4	7	10
N_{JR}	3	6	9
N_{JP}	1	1	1

5. For the design with six links ($N_L = 6$) and seven joints ($N_J = 7$), from Figure 4.13(a) and (b), there are two (6, 7) kinematic chains as shown in Figure 2.13.

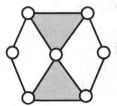 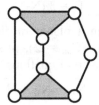

Figure 2.13 Atlas of (6, 7) kinematic chains

6. By assigning one link as the frame (link 1), one link as the swing arm for linking to the tire (link 3), and two binary links in series (dyad) with revolute joints at both ends as the shock absorber (link 5 and link 6) to each (6, 7) kinematic chain shown in Figure 2.13, all possible topological structure of the mechanism for the six-bar planar motorcycle rear suspension can be synthesized. Figure 2.14 shows six of them in which Figure 2.14(b) is the design of the Kawasaki uni-trak, Figure 2.14(c) is that of the Suzuki full-floater, and Figure 2.14(e) is design of the Honda pro-link.

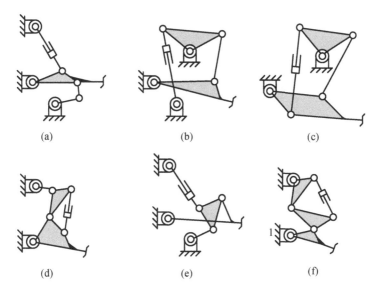

Figure. 2.14 Rear suspension mechanisms of motorcycles

References

1. Yan, H.S., Mechanisms (in Chinese), 3rd edition, Dong Hua Books, Taipei, 2006.
   顏鴻森，機構學，第三版，東華書局，台北，2006 年。
2. Yan, H.S., Creative Design of Mechanical Devices, Springer, Singapore, 1997.
3. Tian Gong Kai Wu (in Chinese) by Song Ying-xing (Ming Dynasty), Taiwan Commercial Press, Taipei, 1983.
   《天工開物》；宋應星[明朝]撰，台灣商務印書館，台北，1983 年。
4. Nong Shu (in Chinese) by Wang Zhen (Yuan Dynasty), Taiwan Commercial Press, Taipei, 1968.
   《農書》；王禎[元朝]撰，台灣商務印書館，台北，1968 年。

Chapter 3 Ancient Chinese Machines

There were numerous mechanical inventions in ancient China, but few people know about these achievements due to the lack of surviving archives. This chapter first introduces the historical development of ancient Chinese machines followed by ancient labor-saving devices. Finally, major mechanical elements and mechanisms of ancient Chinese machines, such as linkage mechanisms, cam mechanisms, gear mechanisms, and flexible connecting mechanisms (ropes and chains), are presented.

3.1 Historical Development

Based on operating principles, the materials used, the power sources given and the design and manufacturing technology available, ancient Chinese machines before the 15th century can be divided into the following three periods [1, 2].

I. Old Stone Age (~400000–500000 BC) to New Stone Age (~2500 BC)

This was the period of primitive society in ancient Chinese history. It started from around the Old Stone Age about ~400,000–500,000 years ago to the New Stone Age some 4,000–5,000 years ago.

As is the case with any ethnic group, the invention and development of machines always begin with simple designs. During that time, the beginning machines were simple devices. Natural materials such as stones, wooden clubs, seashells, and animal bones were hammered, modified, polished, and cut manually, turning them into stone knives and axes. With these simple labor-saving tools, people were able to perform work unachievable with only bare hands. The tools evolved into simple machines such as primitive weaving machines and potter's wheel heads. These simple machines were used in agriculture, fishing, hunting, weaving, and construction work.

In this period, people applied the principles of the lever, wedge and inertia, as well as the elasticity, thermal expansion and contraction of materials.

II. New Stone Age (~2500 BC) to Eastern Zhou Dynasty (~550 BC)

This was the period of slavery in ancient Chinese history, lasting from around the New Stone Age 4,000–5,000 years ago to the Spring-Autumn and Warring Periods of the Eastern Zhou Dynasty around 2,500 years ago.

As needs developed, people combined simple tools with simple machines into complex machines to achieve more intricate objectives; for instance, the scissors are a combination of the wedge and lever. Regarding the material used in machines, in addition to woods, bronzes and irons were also widely used during this period. Furthermore, simple pulley blocks, levers, wheels, winches, and arrows were developed into complex machines such as chariots and weapons. The book Kao Gong Ji 《考工記》 [3] during the Warring Period (480–222 BC) compiled the manufacturing experiences of many kinds of handicraft, and it also showed the manufacturing level of the craftsman at that time.

III. Eastern Zhou Dynasty (~550 BC) to Ming Dynasty (AD 1368–1644)

This was the period of feudal society, starting from the early Warring Period approximately 2,500 years ago until the Ming Dynasty (AD 1368–1644).

During the Qin Dynasty (221–207 BC) and Han Dynasty (206 BC–AD 220), the development of machines in ancient China was at its prime. The technology level of metallurgy, foundry and forgery was high, especially when its rapid development led to the widespread use of metals. The motion and force transmission of some machines applied links, levers, gears, ropes, belts, and chains. Moreover, some machines were even equipped with gear trains and automatic control devices. There were also major innovations in agricultural and weaving machines, as well as in land and sea transportation. It can be seen from the copper horse chariots found inside the Qin Imperial Tomb that the technology of metal working was superlative. The water-driven wind box that appeared in the Eastern Han Dynasty (AD 25–220) was composed of a water wheel, a rope and belt transmission, a link and lever transmission, and a wind box. The wind box included three basic parts of modern machines, namely, the power source, the transmission mechanism, and the working machine. This period also featured outstanding scholars and inventors such as Zhang Heng (張衡), Ma Jun (馬鈞), Zu Chong-zhi (祖沖之), Yan Su (燕肅), Wu De-ren (吳德仁), Su Song (蘇頌), and Guo Shou-jing (郭守敬). They contributed immensely to the development of machines in ancient China.

Except for weaponry and shipbuilding, there were but few significant developments in mechanical technology during the 100 years between the Ming Dynasty (AD 1368–1644) and the Opium War (AD 1842). Noteworthy is a book named Tian Gong Kai Wu 《天工開物》 by Song Ying-xing

(宋應星) during the 17th century [4]. It was a compilation of production experiences that had existed for a long time in ancient China. The book is a major historical encyclopedia of ancient Chinese technologies.

3.2 Labor-saving Devices

People are different from animals because they have the ability to make tools. Labor was the only force possessed by humans in ancient times, but brute force was enhanced by the use of tools. In prehistoric eras, people were not physically strong competitors against animals; nonetheless, people used their ingenuity to make different tools to secure safety, food, and clothing, as well as improve their shelters.

Ancient Chinese devices and machines were immensely intricate. The main functions were to transmit motions and forces to produce work and transform energies. Yet most of the even very ancient designs were used to save effort during work. The wedge, inclined plane, screw, lever, and pulley were simple devices used to save effort during work. The ancient Chinese referred to them as the five labor-saving jewels. The wedge, inclined plane, and screw were closely related; likewise, were the lever and the pulley.

3.2.1 Wedge

There are many references to the applications of the principle of the wedge in ancient records [5, 6]. The more ancient ones include:
1. Gu Jin Shi Wu Kao 《古今事物考》 [7]
   Shen Nong made the ax. 『神農作斧斤。』 (3218–3079 BC)
2. Gu Shi Kao 《古史考》 [8]
   Scissors are devices made of iron that are used to cut textile. Scissors originated in the time of the Yellow Emperor. 『剪，鐵器也，用以裁布帛，始於黃帝時。』 (2698–2599 BC)
   Gong Shu-ban made the shovel. 『公輸般作鏟。』 (~530 BC)
3. Shi Wu Gan Zhu 《事務紺珠》 [9]
   The chisel that is used to perforate wood was created by Xuan Yuan (the Yellow Emperor). 『鑿，所以穿木，軒轅制。』 (2698–2599 BC)
   The planer, a device for smoothening wood, was created by Lu Ban. 『推鉋，平木器，魯般作。』 (~530 BC)
4. Shuo Yuan 《說苑》 [10]

Confucius heard Wu Qiu-zi holding a sickle and crying.『孔子聞吾丘
子振鎌而哭。』(~530 BC)

The ax, scissors, chisel, shovel, planer, and sickle were applications of the principle of the wedge. In the records of ancient Chinese literature, Shen Nong (神農) was regarded as the first to apply such a principle.

Also, when the Mohists (墨家) during the Eastern Zhou Dynasty (770–222 BC) discussed the labor-saving functions of the wedge, they used the "awl" as a description. Wang Chong (王充) during the Han Dynasty (206 BC–AD 220) said [11]: "The awl can pierce easily. If the blunt end of the awl is used, however, it cannot penetrate even a little bit."『針錐所穿，無不暢達；使針錐末方，穿物無一分之深矣。』Although the theory of mechanics might not have existed in ancient China, it is apparent from these references that the Mohists and Wang Chong understood the resourceful applications of the wedge.

In addition to historical records on the applications of the wedge in ancient China, there are also many pieces of archeological evidence. The wedge was already in use in ancient China during the Paleolithic Period some 400,000 to 500,000 years ago. Small two-edged stone chopping tools were discovered at Zhoukoudian, while larger stone chopping tools and pointed-tip tools were discovered in county Ruicheng of province Shanxi. These stone tools belonged to a period earlier than that of the Peking Man. Furthermore, the stone implements found in Zhoukoudian, site of the Peking Man remains, showed traces of wear and tear as a result of chopping trees. Awls made from bones hundreds of thousands of years old were found in city Ziyang of province Sichuan, Figure 3.1. From the mountain caves in Zhoukoudian, stone implements, bone implements, and needles made from bone tens of thousands of years old were excavated. In the vast areas in northern China, south of the Yangzi River, and northeastern China, large quantities of polished implements made from stones, bones, horns, and animal fangs ranging from 10,000 to 5,000 years old were discovered. During the New Stone Age around 5,000 years ago, there were more types of wedge and awl-shaped devices that were made with better techniques. The use of metallic implements started in ancient China during the Bronze Age over 4,000 years ago. Apart from wedge-shaped bronze implements, there were also metallic implements such as knives, axes, sickles, woks, swords, arrowheads, and ploughs.

The above explanation shows that in ancient China during the prehistoric period from the Stone Age to 2500 BC, natural materials such as stones, wood, and bones were already used to make implements that applied the principle of the wedge, such as stone axes, stone knives, bone needles, fish hooks, fish spears, weaving needles, hoes, and arrowheads.

These implements were used for self-defense and hunting. During the serf-dom period (2500–550 BC starting from the Bronze Age some 4,000 years ago), ancient Chinese applied the principle of the wedge to make different weapons and daily implements using bronze and metal. In addition to many wedge-shaped bronze implements, there were also many metallic implements such as knives, axes, sickles, woks, shovels, swords, arrow-heads, and ploughs. The principle of the wedge had been applied to make different weapons and daily implements using bronze and metal since the start of the feudal period (550 BC), or since the Warring Period over 2,500 years ago. The application of the principle of the wedge may be found in production machines, daily implements, or military weapons. Some of them were used together with the lever to make scissors and hoes.

The knives, cutting tools, and nails in our daily lives today also use the principle of the wedge.

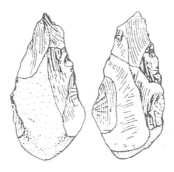

Figure 3.1 Stone chopping and pointed-tip tools in the Old Stone Age [1]

3.2.2 Inclined plane

Although ancient Chinese knew how to use the inclined plane long ago, there were only a few ancient records of inventions using the principle of the inclined plane [5, 12].

1. Kao Gong Ji 《考工記》 [3]
   More work required in going up a hill. 『登阤者，倍任者也。』 (~550 BC)
2. Xun Zi·You Zuo 《荀子·宥坐》 [13]
   One cannot lift an empty cart to a plane three chi (尺, ancient Chinese length of foot) high, but a fully loaded cart can go up a mountain. This is because the mountain path is inclined. 『三尺之岸而虛車不能舉也，百仞之山任負車登焉。何則？陵遲故也。』

Since the inclined plane applies a labor-saving principle, it is mainly used in the transmission of motion or application of force. For this reason, there were no specific inclined plane inventions and no unique archeological findings. It is generally believed that the inclined plane was used as early as the Paleolithic Period.

The Mohism School founded by Modi (墨翟) constructed a gravity-pull cart to demonstrate the effects of the inclined plane. Figure 3.2 shows such a concept [14]. The front wheel was smaller than the rear wheel. A flat board was placed on the wheels, which made the board inclined. A rope tied to the rear wheel axle was passed through a pulley installed atop the inclined board. A heavy object was tied to the other end of the rope. This way, only a small force was needed to push the cart forward, and heavy loads could be lifted to specific heights. Although the gravity-pull cart might be a hard-earned scientific breakthrough for the Mohists, it is common sense for today's movers.

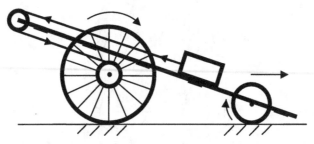

Figure 3.2 Concept of Modi's gravity-pull cart of the inclined plane [14]

Experiences and applications of the principle of the inclined plane to conserve energy could be seen in day-to-day activities in ancient China. For instance, using an inclined ladder to climb higher, using inclined stairways to go up buildings and towers, and using a winding path to go up mountains were common practices. Nevertheless, ancient literature con-tained few references to the inclined plane.

In today's daily life, the inclined plane is a necessary application every-where, such as stairs for going up floors and slopes for going up mountains.

3.2.3 Screw

Unfortunately, ancient Chinese literature does not have adequate records on the invention and application of the screw before the Ming Dynasty (AD 1368–1644) [05, 15].

1. Bao Pu Zi · Chapter 15 of Nei Pian《抱朴子·內篇卷十五》[16]

   [O]r use wooden propeller, cow hide wrapped around the shaft of a flying device ... rises forty li (里, ancient Chinese distance of mile) up ...

   『... 或用棗心木爲飛車，以牛革結環，劍以弔其機， ... 上升四十里，... 』

   This describes a flying device named fei che (飛車) that used the principle of the screw. Recently, Liu Xian-zhou (劉仙洲) studied and reproduced such a mechanism, Figure 3.3 [17].

2. San Cai Tu Hui Chapter 8 of Qi Yong《三才圖會·器用八卷》[18] (AD 1609)

   Figure 3.4 shows a figure of a section of a musket, which used the principle of the screw in weaponry.

A clear description of the invention and application of the screw in ancient literature appeared in the books Wu Bei Zhi《武備志》(AD 1621), Pictorial Book of Strange Devices of the Far West《遠西奇器圖說》(AD 1627), and Wu Li Xiao Shi《物理小識》(AD 1640) during the Ming Dynasty. Nevertheless, these undoubtedly had been influenced by the introduction of western technology following Mateo Ricci's travel to China in AD 1600.

Because there were no records of inventions and applications of the screw before the Ming Dynasty (AD 1368–1644), there were also no artifacts found. The "bamboo dragonfly," a children's toy in ancient China, used the principle of the screw; likewise, the "dragon tail pump" (Archimedean screw) that was used to draw water in earlier times.

Since the screw is a simple labor-saving mechanism, it is widely used today. For instance, the translation screw or power screw is often used to transform circular motion to linear motion, and it is often used in transport mechanisms, production machines, machine tools, spinning machines, weapons, and antenna systems.

3.2.4 Lever

The shadoof and the weighing balance are applications of the principle of the lever [5, 19].

Shadoof

The application of a lever in ancient China was recorded approximately in 1700 BC. Yi Yin (伊尹), a minister during the Shang Dynasty (1766–1122 BC), invented the shadoof namely jie gao (桔槔) which was used for irrigation and for drawing water.

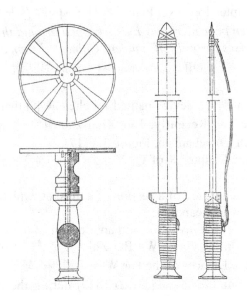

Figure 3.3 Reconstruction of fei che (飛車) [17]

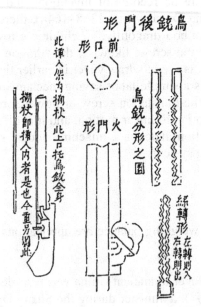

Figure 3.4 Screws in San Cai Tu Hui《三才圖會》[18]

The shadoof was a device that used the lever principle to draw water. A big tree or stand beside the well was made the support, that is, the frame, and a horizontal rod was placed on the support. A water pail for lowering into the well was attached to a hook on a vertical long rod connected to one end of the horizontal rod, while a rock was tied to the other end of the horizontal rod to balance the weight, Figure 3.5 [20].

Figure 3.5 An ancient shadoof [20]

The earliest record of the shadoof appeared in the book Zhuang Zi 《莊子》 [21].

1. Zhuang Zi·Chapter 12 of Wai Pian Tian Di 《莊子·外篇天地第十二》

Zi Gong traveled south to the State of Chu. When he passed by Hanyin, he saw a man in the vegetation who dug a well and went inside it carrying a water jar. The man spent a lot of energy but accomplished little work. Zi Gong said, 'There is a device that irrigates a vast area of vegetation. Little effort is needed but the accomplishment is enormous. Why not try it?' The man asked about the device, and Zi Gong said, 'Make a mechanism using wood, and make the rear heavier than the front. The device can draw so much water, and it is called the shadoof.' 『子貢南遊于楚，反于晉。過漢陰，見一丈人，方將爲圃畦。鑿隧而入井，抱甕而出灌。搰搰然用力甚多而見功寡。子貢曰，有械于此，一日浸百畦，用力甚寡而見功多，夫子不欲乎？爲圃者仰而視之曰：奈何？曰：鑿木爲機，后重前輕，挈水若抽，數如泆湯，其名爲桔槔。』

2. Zhuang Zi · Chapter 14 of Tian Yun《莊子 · 天運篇十四》

   Yan Yuan asked Shi Jin, 'Have you not seen a shadoof? If you pull, it comes down. If you let go, it goes up.' 『顏淵問師金曰：子獨不見桔槔者？乎引之則俯，舍之則仰。』

In the publication Qi Min Yao Shu《齊民要術》[22] and the book Nong Zheng Quan Shu《農政全書》[20], the shadoof was a primary water irrigating mechanism.

Weighing balance

Another application of the lever in ancient China was the weighing balance (權衡).

A hanging rope was attached to the lever to serve as support. A heavy object was attached to one end of the lever while a measuring weight was hung on the other end to measure the weight of the object. This device, referred to as the weighing balance by ancient people, was composed of the measuring weight and the balancing lever.

There are many records of inventions relevant to the weighing balance in ancient literature. The weighing balance was a common application during the Spring–Autumn and Warring Periods (770–222 BC). Early records include:

1. Lu Shi Chun Qiu《呂氏春秋》[23]

   The Yellow Emperor instructed Ling Lun to gather bamboos in the valley of Kun Lun to make a musical instrument; a balance was made to weigh them. 『黃帝使伶倫取竹于崑崙之山嶰谷，爲黃鐘之律，而造權衡度量。』(2599–698 BC)

2. Mohist Canon · Chapter 43 of Shuo Xia《墨經 · 說下第四十三》[24]

   The balance tilts if the weight is added to one side; this is due to matching of the weights. Level both sides and the base becomes shorter and the tip longer. Add equal weights on both sides and the tip goes down; this is due to the tip having gained weight. 『衡，加重于其一旁必捶，權重相若也。相衡，則本短標長。兩加焉，重相若，則標必下，標得權也。』

3. Zhuang Zi · Chapter 10 of Wai Pian Qu Qie《莊子 · 外篇胠篋第十》[21]

[I]t is called the weighing balance ... the balance that measures weights prevents disputes between the people ... 『 ... 爲之權衡以稱之 ... , ... 掊斗折衡而民不爭 ... 』

4. Mencius《孟子》[25]

[W]ith the weighing balance weighs are determined ... 『 ... 權然后知輕重 ... 』

The book Mohist Canon《墨經》was the earliest to explain the principle of the lever in the weighing balance. It referred to the section from the fulcrum to the tip where the heavy object was placed (arm of resistance) as ben (本), while the portion from the fulcrum along the measuring scale (arm of effort) as biao (標). Mohists discussed the different settings of the weighing balance, which included the equal and unequal balancing of the ben and biao against the distances between the force and effect. Two hundred years before Archimedes discovered the principle of the lever, the ancient Chinese had already invented the weighing balance with two fulcrums, namely the zhu ping (銖秤). In using the zhu ping, the fulcrum, not the weighted lever, needed to be adjusted to measure the weight of an object.

In addition to historical records, there are also many artifacts that attest to the use of lever devices in ancient China. When man started to use coarse stone knives and axes, they might already have known how to use easily obtainable wooden clubs or sticks. Because wooden tools could not survive for a long time, they were difficult to preserve. In those times, most of the stone knives and axes with holes had wooden handles, which served as levers when the tools were used. Today, the earliest weighing balance excavated was the wooden balance with bronze weights (木衡銅權), Figure 3.6 [5], excavated from a tomb of the State of Chu of the Warring Period (480–222 BC) in mountain Zuojiagong near city Changsha. It was a device belonging to the 3rd or 4th century BC. Furthermore, one of the paintings of the 28 Heavenly Gods by Zhang Seng-yao (張僧繇) during the Liang Period of the Southern and Northern Dynasties (AD 386–589) showed the use of a wooden balance with bronze weights and weights for weighing objects. Indeed, ancient China bore many artifacts of weighted levers. In addition to the weighted lever mentioned in the book Nong Zheng Quan Shu, images of the weighted lever also appeared in wall paintings in the Wu Liang Shrine of the Han Dynasty (206 BC– AD 220), and in paintings of Jiao Bing-zhen (焦秉貞) during the Qing Dynasty (AD 1644–1911). In addition to the shadoof and weighing balance, there were also many applications of the principle of the lever in ancient China. These included scissors, hoes, pincers, pedal-driven trip hammers, water-driven trip hammers, catapults, and the pedals of spinning machines, and others.

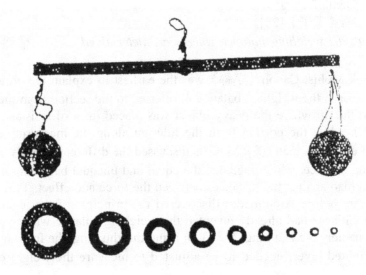

Figure 3.6 An ancient wooden balance with bronze weights [5]

The lever is used either directly or in combination with other simple mechanisms. Primitive people used clubs to move heavy rocks or battle with wild beasts; this was a direct application of the lever. The stone knives and axes used by Stone Age people were tied to wooden poles, or the wooden poles were inserted through the holes of their stone implements, these by making combinational use of the lever and the wedge.

Today, the bottle opener, weight, balance, nail-extracting hammer, and even the badminton racket are applications of the principle of the lever. On the other hand, the nail cutter, scissors, and pliers are examples of the combined application of the principles of the lever and the wedge.

3.2.5 Pulley

The ancient Chinese knew how to use the pulley very early [5, 26]. Important records on inventions based on the principle of the pulley include:
1. Wu Yuan《物原》[27]
   Shih Yi was the first to create the pulley block. 『史佚始做轆轤』
   Shih Yi (史佚) was a historian during the early Zhou Dynasty. If the record is credible, the pulley block was invented sometime around 1100 BC.

2. Nong Shu 《農書》 [28]

From the record in the book Nong Shu, the dual-way pulley block was invented before AD 1313.

3. Jin Shi · Chapter 107 《晉史·卷一百七》 [29]

On the Shizigang of the wetern city of Handan there is the grave of Yue Jian-zi. Ji Long ordered the grave excavated. The first layer was composed of coal more than one zhang (丈, ancient Chinese length of ten feet) deep. The second layer was a wood panel one chi thick. When the panel was raised, it was discovered that there was a spring with clear and cold water eight chi below the panel. A jug made from cowhide was attached to a winch cart to draw the spring water. Because the spring could not be drained in over a month, the excavation was discontinued.
『邯鄲城西石子堈上有越簡子墓，至是季龍令發之。初得炭深丈餘，次得木板厚一尺。稱板厚八尺乃及泉，其水清泠非常。作絞車以牛皮囊汲之，月餘而水不盡，不可發而止。』(AD 336)

4. Wu Jing Zong Yao 《武經總要》 [30]

The winch cart uses large lumber as base. Two cross beams are installed in the front where a winch crank is placed. Four wheels are installed below. The mechanism is massive, and it can move an object weighted one thousand jin (斤, ancient Chinese weight of pound). 『絞車，合大木爲床，前建二叉手柱，上爲絞車，下施四單輪，皆極壯大，力可挽二千斤。』

The book Mohist Canon was the earliest to discuss the theory of the mechanics of pulleys. It described the force in hoisting the heavy object as qing (擎), the force in free fall as shou (收), and the entire pulley contraption as sheng zhi (繩制, rope measure). The Mohist Canon also described using the rope to raise heavy objects, where the forces of qing and shou were opposing but their effects existed on the same point. Raising heavy objects qing required work, but no effort was needed in shou. Using the "rope measure," one could save energy. On one side of the "rope measure," the rope was longer, the object was heavier, and the load tended to fall even lower. On the other side, the rope was shorter, the object lighter, and the load tended to rise higher. If the rope became vertical, then the weights of the objects were equal, and the "rope measure" was in equilibrium. If the "rope measure" was not in equilibrium, the heavy object being raised must be on the sloped side rather than hanging freely in midair.

The pulley was invented in ancient China very early, and its application was very common. Chapter 12 of Shuo Xia of Mohist Canon 《墨經·說下第十二》 [24] discussed experiments of the pulley. Lu Ban (魯般) of the Warring Period (480–222 BC) used pulleys to bury the mother of Ji

Kang-zi (季康子), and built scaling ladders for the State of Chu to attack the State of Song. Also, there is a stone image in the Wu Liang Temple in province Shandong that depicts the story of the First Emperor of Qin Dynasty finding a cooking vessel in river Si Shui, Figure 3.7 [31]. According to legend, Da Yu (大禹) made nine cooking vessels for distinguishing good and evil. The implements became the symbol of power of emperors during the Xia Dynasty (2205–1766 BC), Shang Dynasty (1766–1122 BC), and Zhou Dynasty (1122–256 BC). During the 19th year of Zhou Nan Wang (周赧王) (296 BC), Qin Zhao Wang (秦昭王) took the nine cooking vessels with him and lost one in river Si Shui. When the First Emperor of Qin passed by the place on his way to the Eastern Sea in search for eternal life, he ordered 1,000 men to go into the river to retrieve the lost cooking vessel. When the cooking vessel appeared on the river surface, a dragon came out and gnawed on the rope, broke it, and caused the implement to sink into the river bottom again. The image of the vessel being salvaged depicts three men bending their body and exerting effort to pull ropes on the slopes of both banks of the river. One end of the ropes was passed through a pulley and attached to the top of the cooking vessel. There was an audience in the scene. This adequately proved that pulleys were commonly used at the time.

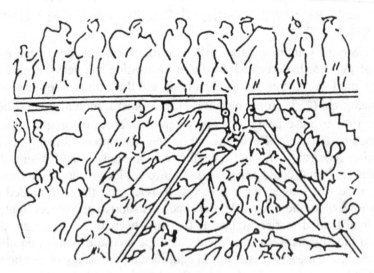

Figure 3.7 Pulley in an ancient stone image [31]

Single-way pulley block

Another form of pulley was a pulley block (轆轤), which was used to draw water from wells, Figure 3.8 [4]. It was composed of a short wooden cylinder placed across supports beside the well. A rope was coiled around the cylinder. One end of the rope was permanently fixed on the cylinder, while the other end was tied to a water pail. A crank was used to rotate the wood cylinder to draw water. This was called the single-way pulley block. Because the radius of rotation of the crank was greater than the radius of the cylinder, the effect of force magnification could be produced.

Figure 3.8 An ancient single-way pulley block [4]

Due to needs and changes in applications, the single-way pulley block was developed into a dual-way pulley block, winch cart, differential pulley, compound pulley, and axle. These are explained below.

Dual-way pulley block

The dual-way pulley block was also referred to as the flowering pulley block or compound pulley block, with two ropes coiled in opposing directions. The ends of the ropes were tied to water pails. When the full water pail was pulled upward, the empty water pail would go down. Compared with the single-way pulley block, the dual-way pulley block was functional in either direction. Moreover, the weight of the empty water pail and the rope going down the well minimized the effort needed to do the work. Therefore, the diameter of the pulley block could be increased to save time when drawing water.

Winch cart

The winch cart was an offshoot of the development of the pulley block. Compared with the pulley block, it had longer curved arms for exerting force. Also, the winch cart had many curved arms.

Differential pulley

The differential pulley, known to Westerners as the Chinese windlass, is composed of two rounded shafts with different diameters. The pulley where the heavy object is attached is pulled by a rope. One end of the rope is coiled around the shaft with the bigger diameter, while the other end is coiled around the shaft with the smaller diameter. One half the differences in distance between the rise and fall of both ends is generally equivalent to the distance traveled by the heavy object. If the difference between the diameters of the shafts is very small, then the distance traveled by the heavy object is also very small when the pulley makes one rotation, thus magnifying the effect of the force.

Compound pulley

A compound pulley is composed of two pulleys with different diameters attached to an axle. Force is applied to turn the larger pulley, while the smaller pulley moves the heavy object. Its main effect is to reduce the speed of movement of the last moving pulley. This way, a smaller force can be used to move a heavy object.

3.3 Linkage Mechanisms

A linkage mechanism is an assemblage of links to transform types and directions of motions, coordinate required state of motions, guide rigid bodies, or generate motion paths [32].

Ancient China has a very long history in the use of links (lian gan) and linkage mechanisms, but their exact dates of use cannot be validated from the literature and artifacts. Furthermore, the term lian gan (連桿, link) is seldom seen in ancient manuscripts; instead, qu bing (曲柄, crank), gang gan (槓桿, lever), or hua jian (滑件, slider) are common. They all refer to today's "link."

The earliest application of the link appeared in the Old Stone Age. In the beginning, it was only a simple crank or lever, but later, levers were interconnected to form link-lever mechanisms to boost work efficiency. The development of links and link-lever mechanisms was from simple to complex; it showed the close relationship of daily life and machines, such as agricultural machines, weaving machines, and handicraft machines.

In what follows, the historical records, excavated proofs, and application examples of ancient Chinese linkage mechanisms are presented.

Man-powered mill

A mill is a device for removing rice hulls. The books Nong Shu [28], Nong Zheng Quan Shu [20], and Tian Gong Kai Wu [4] have records and illustrations of the machine.

The man-powered mill was a spatial linkage mechanism. It was an assembly of two ropes (member 2, K_{R1} and K_{R2}), a horizontal rod (K_{L1}), a connecting link (K_{L2}), a crank (K_{C1}), and the grinding stone (K_{C2}), Figure 3.9(a). One end of each rope was secured to the three-legged rack (member 1, K_F); the other end was attached to the horizontal rod. The connecting link and the horizontal rod move parallel with the ground as one assembly (member 3), thereby driving the crank to rotate. The crank and the grinding stone on top of the mill was another assembly (member 4). During operation, the operator used both hands to push the horizontal rod back and forth with slight swaying. This caused the mill to operate continuously with the linkage mechanism to remove the rice husks.

Since the two ropes are designed for providing an efficient input through human power and are symmetric, this device can be analyzed as a spatial mechanism with four members (the ground link K_F, member 1; the rope K_R, member 2; the horizontal and connecting rod K_L, member 3; and the crank and the grinding stone K_C, member 4) and four joints consisting of two spherical joints (J_S; joint *a* and joint *b*) and two revolute joints (J_R; joint *c* and joint *d*). Figure 3.9(b) shows its corresponding schematic drawing. Therefore, the number of members is $N_L = 4$, the number of degrees of constraint of a revolute joint is $C_{sR} = 5$, the number of revolute joints is $N_{JR} = 2$, the number of degrees of constraint of a spatial joint is $C_{sS} = 3$, and the number of spatial joints is $N_{JS} = 2$. Based on Equation (2.2), the number of degrees of freedom, F_s, of this mechanism is:

$$F_s = 6(N_L - 1) - (N_{JR}C_{sR} + N_{JS}C_{sS})$$
$$= (6)(4 - 1) - [(2)(5) + (2)(3)]$$
$$= 18 - 16$$
$$= 2$$

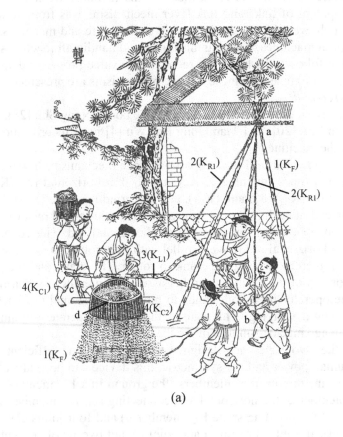

(a)

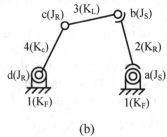

(b)

Figure 3.9 A man-powered mill [4]

Foot-operated spinner
The best application of linkage mechanisms in ancient China was the spinning machines with an intricate assembly of links and pedals. The foot-operated spinner was developed based on the hand-operated spinner. The book Nong Shu [28] recorded the use of the foot-operated spinner with cotton thread rack. This foot-operated spinner was also a type of linkage mechanism, Figure 3.10(a). It was an assembly of a fixed pivot rod (member 1, K_F), a pedal (member 2, K_{L1}), a crank (member 3, K_{L2}), and a large rope wheel (member 3', K_{L3}). One end of the pivot rod was secured to the frame of the machine (member 1, K_F), while the other end was connected to the foot pedal through a spherical joint (J_S; joint *a*). The other end of the foot pedal was connected to the crank by another spherical joint (J_S; joint *c*). The crank and the large rope wheel were one assembly, and was connected to the crank using a revolute joint (J_R; joint *b*). Figure 3.10(b) shows its corresponding schematic drawing. Alternately, stepping on the two sides of the foot pedal caused the crank to rotate the big rope wheel, and the machine to spin.

This is a spatial three-bar mechanism with three joints, where the number of members is $N_L = 3$, the number of degrees of constraint of a cylindrical joint is $C_{sC} = 4$, the number of cylindrical joints is $N_{JC} = 1$, the number of degrees of constraint of a spatial joint is $C_{sS} = 3$, and the number of spatial joints is $N_{JS} = 2$. Based on Equation (2.2), the number of degrees of freedom, F_s, of this mechanism is:

$$\begin{aligned} F_s &= 6(N_L - 1) - (N_{JC}C_{sC} + N_{JS}C_{sS}) \\ &= (6)(3 - 1) - [(1)(4) + (2)(3)] \\ &= 12 - 10 \\ &= 2 \end{aligned}$$

Water-driven wind box
A wind box is a blower using water as its power source to generate the oscillating motion of the output fan through linkage transmissions. The earliest record of the water-driven wind box is found in the Du Shi Zhuan of the book Hou Han Shu《後漢書·杜詩傳》[33], which claims that in the 7th year of Jian Wu (AD 31) during the Eastern Han Dynasty (AD 25–220), officer Du Shi of city Nanyang in province Henan, "constructed the water-driven wind box … doing more work with lesser effort. It was a convenience to the people."『造作水排 … 用力少而見功多，百姓便之。』 Figure 3.11(a) shows the illustration of a horizontal-wheel water-driven wind box in the book Nong Shu [28]. However, the topological structure of this illustration is not clear, and even confusing. This is one difficulty encountered in the study of ancient machines.

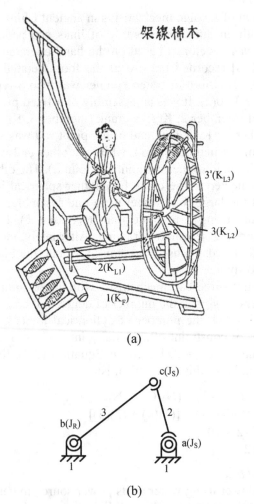

架線棉木

3'(K_{L3})

3(K_{L2})

2(K_{L1})

1(K_F)

(a)

c(J_S)

3

2

b(J_R)

a(J_S)

1   1

(b)

Figure 3.10 A cotton thread rack [28]

From the viewpoint of kinematics, this design consists of three linkage mechanisms in series: a simple rope and pulley mechanism, a spatial four-bar linkage, and a planar four-bar linkage, Figure 3.11(b).

The river water drives the rope and pulley mechanism, which includes a water wheel (member 2, K_{W1}), a vertical axis (member 2, K_{Wa}), a big pulley (member 2, K_{W2}), a small pulley (member 4, K_{W4}), and a rope (member 3, K_R). The water wheel, the vertical axis, and the big pulley have no relative motion to each other, and they are incident to the frame (member 1, K_F) with a revolute joint (J_R; joint a). Through the belt, the big pulley speeds up the small pulley, which is incident to the frame with a revolute joint (J_R; joint b).

Figure 3.11 A water-driven wind box [1, 28]

The spatial four-bar mechanism includes a crank (member 4', $K_{W4'}$) which is integrated to the small pulley, a connecting link (member 5, K_{L5}), and a rocker (member 6, K_{L6}). The crank has no relative motion to the small pulley and is also incident to the frame with a revolute joint (J_R; joint b). The connecting link is incident to the crank with a spherical joint (J_S; joint c) and to the rocker with another spherical joint (J_S; joint d). The other end of the rocker is incident to the frame with a revolute joint (J_R; joint e). This spatial mechanism has four members (members 1, 4, 5, and 6) and four joints consisting of two revolute joints (b and e) and two spherical joints (c and d). Therefore, the number of members is $N_L = 4$, the number of degrees of constraint of a revolute joint is $C_{sR} = 5$, the number of revolute joints is $N_{JR} = 2$, the number of spatial joints $C_{sS} = 3$, and the number of degrees of constraint of a spatial joint is $N_{JS} = 2$. Based on Equation (2.2), the number of degrees of freedom, F_s, of this mechanism is:

$$F_s = 6(N_L - 1) - (N_{JR}C_{sR} + N_{JS}C_{sS})$$
$$= (6)(4-1) - [(2)(5) + (2)(3)]$$
$$= 18 - 16$$
$$= 2$$

The motion of this mechanism is constrained since the rotation of member 5 about the axis through the centers of spherical joints c and d has an extra degree of freedom that does not affect the input (the crank) and the output (the interrocker) relation of this spatial four-bar mechanism.

The planar four-bar linkage includes the rocker (member 6, K_{L6}) as the input, a final connecting link (member 7, K_{L7}), and the output fan (member 8, K_{L8}) of the wind box machine. The final connecting link is incident to the rocker with a revolute joint (J_R; joint f) and to the output fan with another revolute joint (J_R; joint g). The rocker is also incident to the frame with a revolute joint (J_R; joint h). This planar mechanism has four members (members 1, 6, 7, and 8) and four revolute joints (e, f, g, and h). Therefore, the number of members is $N_L = 4$, the number of degrees of constraint of a revolute joint is $C_{pR} = 2$, and the number of revolute joints is $N_{JR} = 4$. Based on Equation (2.1), the number of degrees of freedom, F_p, of this mechanism is:

$$F_p = 3(N_L - 1) - (N_{JR}C_{pR})$$
$$= (3)(4 - 1) - (4)(2)$$
$$= 9 - 8$$
$$= 1$$

Therefore, the motion of this mechanism is constrained.

In summary, this complex linkage mechanism has one input (the water wheel) and one output (the oscillating fan) and has constrained motion.

Jie Chi

In ancient China, Jie chi (界尺) was a traditional art tool used for drawing parallel lines, Figure 2.1(b). It was composed of upper and lower rulers of equal lengths joined by hinges to two bronze plate levers of equal lengths. When the lower ruler was set, moving the bronze lever and the angle of the vertical ruler would cause the upper ruler to become parallel to the lower ruler. It is a parallelogram consisting of four members (members 1, 2, 3, and 4) with four revolute joints (J_R; joints *a*, *b*, *c*, and *d*). The schematic representation of this device is shown in Figure 3.12.

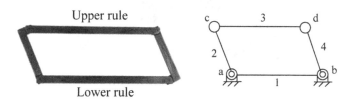

Figure 3.12 Schematic representation of jie chi (界尺)

3.4 Cam Mechanisms

A simple cam mechanism consists of three basic parts, a cam, a follower, and a frame. A cam is an irregularly shaped machine member which serves as a driving link by rotating with a constant velocity and imparting motion through direct contact to a driven link, the follower, which in turn moves in a desired motion. A cam (K_A) is adjacent to a follower (K_{Af}) and a frame (K_F) with a cam joint (J_A) and a revolute joint (J_R), respectively. A follower, which is adjacent to the frame with a revolute joint or a prismatic joint (J_P), is usually driven to move with varying speeds in a noncontinuous and irregular motion. Figure 3.13 shows a simple cam mechanism together with its kinematic chain.

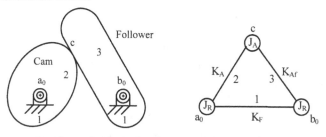

Figure 3.13 A simple cam mechanism [32]

Cam mechanisms had been in use for a long time in ancient China. The trigger mechanism of the crossbow around 200 BC was an intricate cam-shaped swing arm. The invention of the cam may be traced back to the Spring–Autumn and Warring Periods (770–222 BC), specifically to the bronze crossbow (Figure 3.14) belonging to the Warring Period (480–222 BC), which was excavated in county Yi of province Hebei.

Figure 3.14 An excavated bronze crossbow [1]

Cam mechanisms also appeared in water-driven pestles. The publication Huan Zi Xin Lun 《桓子新論》 during the latter part of the West Han Dynasty (206 BC–AD 8) records a water-driven tilt hammer that "... used water to pound ..." 『... 役水而舂 ...』[34].

Complex water-driven multiple tilt hammers appeared as early as the Jin Dynasty (AD 265–420). The publication Jin Zhu Gong Zan《晉諸公讚》 states: "Du Yu and Yuan Kai constructed water-driven multiple tilt hammers." 『杜預元凱作連機水碓』[35] There are also many records of water-driven multiple tilt hammers in the literature of the later periods.

Figure 3.15(a) shows a water-driven multiple tilt hammer described in the book Tian Gong Kai Wu [4]. This is a typical simple cam mechanism with three members and three joints. Figure 3.15(b) shows its corresponding schematic drawing. The water wheel (member 2, K_W) was connected to a long shaft (member 2, K_{Wa}) with paddles (member 2, K_{Ac}) as an assembly. When the water flow moved the water wheel, its cam effect on the assembly caused the tilt hammers (member 3, K_{Af}) to produce work. The long shaft was incident to the frame (member 1, K_F) with a revolute joint (J_R; joint a). The paddles were incident to one end of the tilt hammers with cam joints (J_A; joint c). The tilt hammers were also incident to the rack through a revolute joint (J_R; joint b).

The bell and gong mechanism on the hodometer (記里鼓車) was also a cam mechanism. The Yu Fu Zhi of the book Song Shi《宋史·與服志》 [36] describes such a mechanism as: "the shaft of the big outer wheels has two metal paddles; the horizontal wooden shaft has a paddle each." 『外大

平輪軸上有鐵撥子二，又木橫軸上有撥子各一。』In the book Nong Shu [28], the description of the transmission of the vertical wheel of the water-driven wind box, which used a cam device, was similar to that of the water-driven tilt hammer.

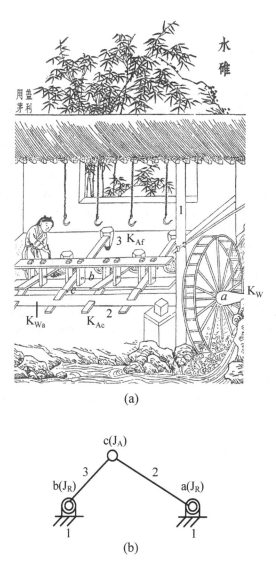

(a)

(b)

Figure 3.15 A water-driven multiple tilt hammer [4]

Others machines like the wood figurine clocker inside the water-driven astronomical devices by monk Yi-Xing (一行) and Liang Ling-zhan (梁令瓚) during the Tang Dynasty (AD 618–906), the spurs inside the water-powered armillary sphere and celestial globe, the automated alarm inside the five-wheel sand-driven clock, and the movement of the paper figures inside the revolving lantern, all employed cam mechanisms.

3.5 Gear Mechanisms

Gears are machine parts, operating in pairs, which transmit motion and force from one rotating shaft to another, by means of successively engaging projections called teeth, for producing constant velocity ratio by direct contact. When two or more gears are in mesh for the purpose of transmitting motion from one shaft to another, the gear set is called a gear train. Figure 3.16(a) shows a simple gear train with one degree of freedom consisting of a driving gear (member 2, K_{G2}), a driven gear (member 3, K_{G3}), and a frame (member 1, K_F). The motion and power of the driving shaft are transmitted directly from the driving gear to the driven gear that in turn drives the driven shaft. The two gears are adjacent by a gear joint (J_G; joint c), and are each adjacent to the frame with a revolute joint (J_R; joint a_0 and joint b_0). Figure 3.16(b) shows its corresponding kinematic chain.

Based on excavated artifacts, the earliest metallic gears in ancient China may be traced back to no later than the Han Dynasty (206 BC–AD 220). There might have been wooden gears in earlier periods, but the materials could have rotted from age.

Although there are many excavated ancient metallic gears, there has never been any record of their appearance or invention in ancient manuscripts. Generally, gears were referred to as ji lun (機輪), lun he ji chi (輪合幾齒), and ya lun (牙輪), as in the examples below:

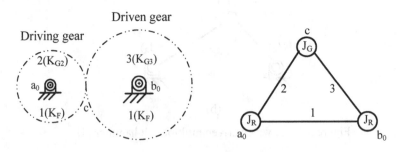

Figure 3.16 A simple gear mechanism [32]

1. Song Shi · Chapter 80 of Lu Li Zhi《宋史·卷八十律曆志》[36]
   [B]elow was a gear wheel with 43 spurs connecting; no human power was necessary ...『 … 其下爲機輪四十有三，鉤鍵交錯相持，不假人力 … 』

2. Yuan Wen Lei · Chapter 5 of Guo Shou-jing Xing Zhuang《元文類·卷五十郭守敬行狀》[37]
   [T]wenty-five small and large wooden gears, where the teeth connected with one another ...『 … 大小機輪凡二十有五，皆以木刻爲沖牙轉相援擊 … 』

3. Ming Shi · Chapter 25 of Tian Wen Zhi《明史·卷二十五天文志》[38]
   In the early Ming Dynasty, Zhan Xi-yuan used a water-driven device but it frozed during winter, so he replaced water with sand ... the five wheels bore 30 teeth ... 『明初詹希元以水漏至嚴寒水凍輒不能行，故以沙代水 … 其五輪惡三十齒 … 』

The term gear did not exist during ancient periods. It appeared only in literature of the late Qing Dynasty (AD 1644–1911). During this time, Chinese mechanical technology was already influenced by the West.

Ancient Chinese gears can be classified based on their functions as power transmission and motion transmission. Gear mechanisms for motion transmissions were primarily used in south-pointing chariots, hodometers, and inside astronomical and clock instruments. However, there has never been any actual item with this type of application passed down or excavated.

Gear mechanisms for power transmissions were primarily used in changing the speed and/or direction of the source power such as human power, animal power, wind power, or water power to produce the required output work. Such designs were often seen in water-driven agricultural machines that did not require high accuracy and speed. Consequently, wood was used as material, and the shape of the teeth was insignificant. They were similar to the pin gears of today.

In what follows, some application examples of ancient Chinese gear mechanisms for power transmissions are presented.

Multiple grinder

Figure 3.17 shows a multiple grinder which appeared during the Jin Dynasty (AD 265–420). It was a food-processing device made up of eight grinders driven simultaneously by a cow. In the device, the cow rotated the large vertical gear wheel, which with its teeth, simultaneously drove the eight smaller horizontal gears.

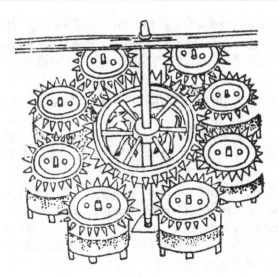

Figure 3.17 A multiple grinder [28]

Water-driven grinder

Figure 3.18 shows a water-driven grinder which was widely used during the Northern and Southern Dynasties (AD 386–589). This water-driven mill was powered by a vertical water wheel. The direction of the power is changed to horizontal by a gear train to drive the output to two mills rotating vertically.

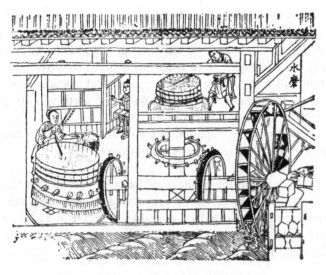

Figure 3.18 A water-driven grinder [28]

Water- and animal-driven mills

Figure 3.19(a) shows a water-driven mill which was driven by a vertical water wheel to produce an output rotation in horizontal direction through a gear mechanism. Figure 3.19(b) shows a horizontal animal-driven mill which was driven by animals to produce an output rotation in the same direction through a simple gear train.

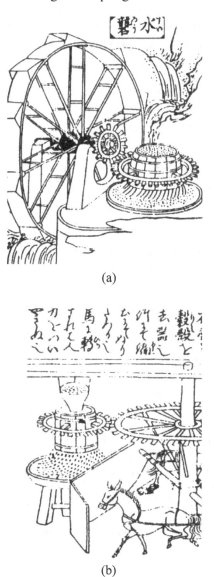

(a)

(b)

Figure 3.19 Water- and animal-driven mills [1]

Cow-driven paddle blade mechanism

The book Nong Shu [28] has detailed introduction of the water-driven and cow-driven paddle blade mechanisms, which used gears to transmit power. Furthermore, the cow-driven paddle blade mechanism can be seen in the Tang Dynasty (AD 618–906) paintings. The books Nong Zheng Quan Shu [20] and Tian Gong Kai Wu [4] also have discussions about it. Figure 3.20 shows a type of cow-driven paddle blade mechanism in the book Nong Shu [28] where the cow is used as the power source to drive the large horizontal gear which connects to the small vertical gear to produce the rotating motion of the horizontal axis for driving the paddle blade mechanism.

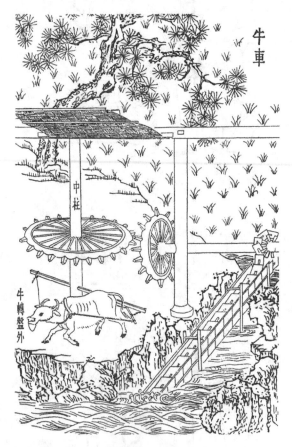

Figure 3.20 A cow-driven paddle blade mechanism [28]

3.6 Rope Drives

Flexible machine elements are used when the distance between the shafts of the driver and the follower is comparatively long. As such, linkage, cam, or gear mechanisms are not applicable. Rather, common types of flexible connecting members are belts, ropes, and chains. The devices that are used for hoisting loads from their tension forces and/or in the transmission of motion and power are called flexible connecting mechanisms. Basically, a flexible connecting mechanism consists of a flexible connecting member that rotates about a pulley, sheave, or sprocket that is fastened to a rotating shaft. Motion and power from the driving shaft are transmitted from the driving wheel (pulley, sheave, or sprocket) through the flexible connecting members to drive the follower (pulley, sheave, or sprocket) which in turn drives the driven shaft. As shown in Figure 3.21, the driving wheel (member 2) and the follower (member 3) are both adjacent to the frame (member 1) with revolute joints (J_R), and are adjacent to the flexible link (member 4) with wrapping joints (J_W).

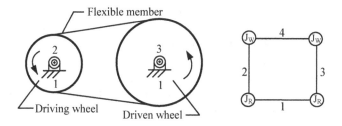

Figure 3.21 A flexible connecting mechanism [32]

Ropes are softer than belts, are easily produced, and can withstand considerable pulling force. Cords are generally used in motion transmission between two nonparallel shafts particularly when the axes of shafts constantly change directions. A common application is in weaving machines. In this case, the sheaves can be rotated in any direction through the pulling of the cords as long as they have sufficiently wide grooves. Wire ropes are suitable for long-distance and high-power rate transmissions, or for long distances with irregular motion paths or power transmission such as in cranes, and in the flight control mechanism of aircrafts.

Ancient China had various applications of flexible connecting mechanisms, especially rope drives and chain drives, in various dynasties. In what follows, the historical development and applications of ancient Chinese rope drives are presented.

In English dictionaries, a rope is defined as a length of stout cord made of strands of natural or artificial fibers twisted or braided together. The term rope also appeared in various ancient Chinese literatures. According to Guang Qi of the book Xiao Er Ya 《小爾雅‧廣器》 [39]: "The larger one is called a rope, the smaller one is called a cord." 『大者謂之索，小者謂之繩。』 According to Xi Ci Xia of the book Yi-Jing 《易經‧繫辭下》 [40]: "Ancient people tied knots on cords to keep record, while people during later periods used writings to keep record." 『上古結繩而治，後世聖人易之以書契。』 And, according to the book Shuo Wen 《說文》 [41]: "The cord was used to tie things together. The cord was made with hemp, while the rope was made with dried weed." 『繩，索也。撚之令緊者也。一曰麻絲曰繩，草謂之索。』

The use of ropes in ancient China can be traced back to the New Stone Age over 4,000 years ago. At that time, ropes would not have functioned as power transmission. During the Shang Dynasty around 1300 BC, ropes for motion and force transmission were used in pulley blocks for drawing water, agricultural machines, and weaving machines. The book Mohist Canon during the Spring-Autumn and Warring Periods (770–222 BC) explored the relationship between the structure and stress of ropes. The rope and pulley used in drilling salt wells appeared later in the West Han Dynasty (206 BC–AD 8) at the latest, but the invention of silk weaving and baste weaving was the earliest in ancient China. Based on excavated wall paintings, it can be proven that cotton spinners and relevant technologies were already available during the Han Dynasty (206 BC–AD 220).

Weaving machines

The evolution of motion and force transmission by ropes in machines was closely related to the development of weaving technology in ancient China. Primitive weaving technology, which was developed from the lashing craft, was already widespread during the New Stone Age. In the beginning, the weaving method involved twisting fibers section by section until a spinning device namely fang zhui (紡墜) was developed. The spinning device could be used to twist and ply. The spinner, which came after the spinning device, was a fully developed weaving machine.

Motion and force transmission by ropes and cords was often seen in the weaving machines of ancient China. In the beginning, the spinner was single-spindle and hand-driven. The main parts were the crank, the wheel, the rope, and the spindle. Figure 3.22 shows a picture of a spinner from the wall painting inside a tomb from the Han Dynasty (206 BC–AD 220). It is a hand-driven single-spindle spinner. The crank is used to rotate the wheel, the rope, and then the spindle shaft. In this way, the spindle can be turned at high speed to spin the thread.

Figure 3.22 A picture of a spinner from the wall painting in the Han Dynasty [1, 5]

The single-spindle spinner was developed into the multiple-spindle spinner, and later, the large spinner. The earliest record of the large spinner can be found in the book Nong Shu [28] as shown in Figure 3.23. The mechanism of a large spinner was divided into two sections. One was for the spindle transmission; the other was for the reel transmission. Bamboo wheels were installed on both sides of the machine, and they were connected by a belt, the pi xian (皮弦) made from animal hide. The lower end of the belt directly weighed down on the spindle rod. When the driving wheel of the left side rotated, friction between the belt and the spindle shaft caused the spindle to rotate. Transmission of the reel was dependent on the effect of a pair of perpendicular wooden wheels and the rope. Friction from the upper end of the belt caused the small rotary drum in the right bottom side to rotate, then the rope caused the rotary drum on top to rotate. The speed ratio of these two rotary drums affected the speed of baste twisting. Based on the transmission from the two sections of rope, the spindle and reel were able to rotate at a specific speed.

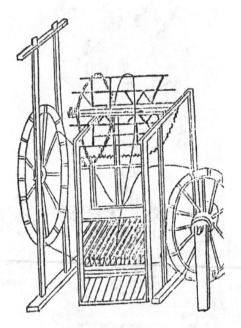

Figure 3.23 A large spinner [28]

Foot-driven silk reeling machine

The main part of a foot-driven silk reeling machine was composed of a linkage mechanism and a rope drive. The book Tian Shu《蚕書》[42] states: "The encircling rope was connected to the left side of the machine. The front measure was five cun (寸, ancient Chinese length of one tenth of a foot). The connecting rod at the bottom left side of the machine was one cun and a half. In the machine was a cylinder, where the encircling rope was coiled. The rope moved continuously as the machine was operated, and it caused the cylinder to rotate. Protruding braces were attached on the cylinder. On the braces were two chi long bamboo rods that receive without tangling the (silk) lines. The crank was attached to the left end of the cylinder, while the other end was placed on the bracket to steady the spool. Therefore, the machine was operated such that the rope rotated the cylinder, and the cylinder caused the bamboo rod to move while maintaining equilibrium."

『車之左端，置環繩。其前尺有五寸，當車床左足之上，連柄長寸有

牛。匠柄爲鼓，鼓生其寅以受環繩。繩應車轉，如環無端，鼓因以
旋。鼓上爲魚，魚半出鼓，其出之中，見柄半寸，上承添梯。添梯
者，二尺五寸片竹也。其上揉竹爲鉤以防系，竅左端以應柄，對鼓爲
耳，方其穿以閑添梯。故車轉以索環繩，繩簇鼓，鼓以舞魚，魚振添
梯，故系不過偏。」Figure 3.24 shows such a design in the book Tian
Gong Kai Wu [4].

Figure 3.24 A foot-driven silk-reeling machine [4]

The wooden cylinder named gu (鼓) (Figure 3.25) in the article was referred to as mu niang deng (牡娘鐙) in the book Guang Tian Sang Shuo Ji
Yao《廣蠶桑說輯要》[43]. At the center of the cylinder was a hole. The
cylinder was placed on a short vertical tenon joint at the left side of the
machine frame. On the outer edge was a sheave for passing the rope. A
rope connected it with the groove on the frame shaft. The upper portion of
the cylinder was passed through the two brackets on short shafts called the
braces. One end of the cylinder protruded beyond the edge of the cylinder.
There was a round tenon on it, making this an eccentric shaft which was
connected to the rod. When the rotating shaft moved, it caused the cylinder
and braces to move too by way of the rope. As a result, the eccentric shaft
on the braces moved in a circular pattern. One end of the spooling bamboo

rod passed through the hole of the pole on the right side of the machine frame. In this way, as the eccentric shaft moved in a circular pattern, the entire bamboo rod moved back and forth. During this time, the bamboo rod fed the silk threads, which intersected and were rolled up in layers.

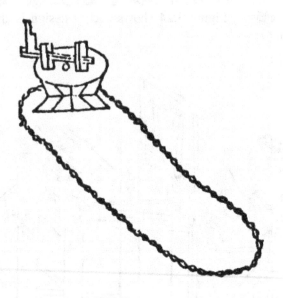

Figure 3.25 *Mu niang deng* (牡娘鐙) - an eccentric device of ropes [1, 43]

This eccentric device of ropes was a typical mechanism in ancient China. The horizontal-wheel paddle blade machine also contained such a mechanism, which transformed the original circular motion into linear reciprocating motion.

Animal-driven mill

In ancient times, animal power was made possible through rope drives. The book Nong Shu states [28]: "The animal force moved the wheel axle, where a belt made from animal hide or a big rope was encircled twice. This was then connected to the upper portion of the mill. As the wheel turned, so did the rope. When the rope achieved one revolution, the wheel had fifteen revolutions. As compared with human labor, this was faster and labor-saving." 『復有畜力輓行大輪軸，以皮弦或大繩繞輪兩周，復交於礱之上級。輪轉則繩轉，繩轉一周則輪轉十五周，比用人工，既速且省。』Figure 3.26 shows such an animal-driven mill from the book Tian Gong Kai Wu [4].

Figure 3.26 An animal-driven mill [4]

Cow-driven well-drilling rope drive (牛轉繩輪鑿井)
The mechanism used in salt-well mining in province Sichuan during the Ming Dynasty (AD 1368–1644) relied on the traction and motion transmission of rope drives. The book Tian Gong Kai Wu states [4]: "A shadoof, pulley block, and other tools were installed at the mouth of the well. A cow was tied and connected to the wheel. As the cow moved, the wheel turned and pulled at the pulley block. Water was then drawn from the well." 『井上懸桔槔、轆轤諸具，置盤架牛，牛拽盤轉，轆轤絞纏，汲水而上。』 Therefore, the cow rotated a big sheave. One end of the rope passed through an idle wheel and pulley block and was tied to the drilling tool. The tool was raised as the cow drove the big rope wheel to rotate. Figure 3.27 shows such a design.

Figure 3.27 A cow-driven well-drilling rope drive [4]

Rope-driven grinding machine
All grinding machines for jade stones in ancient times used ropes or belts for motion and force transmission. Figure 3.28 shows such a design in the book Tien Gong Kai Wu [4]. The grinding wheel is placed on a horizontal axle, and both ends of the axle are attached to bearings. On each side of the grinding wheel is a rope or belt, in which the upper portion is attached to the axle. The ropes or belts are wound in a few circles around the axle in opposite directions. The lower portion of the ropes is attached individually to two foot paddles. Stepping alternately on the foot paddles causes the grinding wheel to rotate back and forth. The jade stone is then brought to the grinding stone for processing.

3.7 Chain Drives

When rigid metal link plates are pinned or hooked together forming a flexible connecting element, it is called a chain. It must be used in conjunction with a sprocket when in transmission to form a chain drive. For long-distance transmissions, gear drives are not economical and belt drives have length deficiency. Chains are thus generally used in such a situation to transmit precise speed ratios.

Figure 3.28 A rope-driven grinding machine [4]

A chain is a hard yet flexible mechanical member for motion and force transmission. There are different designs and shapes of chains depending on the applications. Generally, chains are categorized as hoisting chains, conveyor chains, and power transmission chains. Hoisting chains are used for lifting or towing. A conveyor chain moves the objects hung or placed on it to another place. In addition to moving objects, the conveyor chain is also often used in low-speed force transmission, such as the paddle blade mechanism during the East Han Dynasty (AD 25–220) and the scoop water wheel, named gao zhuan tong che (高轉筒車) during the Tang Dynasty (AD 618–906). The power transmission chain is used for higher speed or higher force transmission, such as the sky ladder named tian ti (天梯) inside the astronomical clock tower during the Northern Song Dynasty (AD 960–1127).

The earliest recorded use of the chain, without the functions of motion and power transmission, is found in the publication General Study of Calabash Instruments in the Shang and Zhou Dynasties (1766–256 BC) 《商周彝器通考》 [44], named Lin Wen Calabash Pot, as shown in Figure 3.29.

Figure 3.29 The chain in the device Lin Wen Calabash Pot (鱗聞瓠壺) [44]

There were many examples of chains used for transporting and conveying purposes within machines in ancient China. Many of them were used in irrigation and water-drawing machines. In what follows, the historical records and application examples of chains and chain drives in ancient China are presented [1].

Paddle blade machine

The paddle blade machine contains a conveyor chain, which allows continuous water-drawing activity. It is convenient to operate and relocate and has been a widely used and effective irrigation or water-drawing machine since ancient times. Based on the source of the power, paddle blade machines can be divided into man-powered, animal-driven, wind-driven, and water-driven. All of them contain the upper and lower chain wheels and conveyor chains as the main parts. The wooden chain of the paddle blade machine is called the "dragon spine." Its main part is referred to as the "crane's knee" in the book Nong Zheng Quan Shu [20], and it is connected by wooden pins to form a chain.

Paddle blade machines had many names, such as dragon-spine machine, water dragon, water machine, foot paddle machine, and water centipede. Most manuscripts in the early periods referred to it as the paddle blade machine. Based on literary records, the paddle blade machine was invented no later than the East Han Dynasty (AD 25–220). The publicationZhang Rang Zhuan of Hou Han Shu 《後漢書·張讓傳》 states [45]: "On the third year of Zhong Ping Period (186 AD), the eunuch Bi Lan was ordered to make four bronze men … and paddle blade machine and siphon on the east side of the bridge for irrigating the streets on the southeastern side, so as to save people from paying for irrigation." 『中平三年又使掖庭令畢嵐鑄銅人四，… 又作翻車，渴烏，施於橋西，用灑南北郊路，以省百姓灑道之費。』 Also, Wei Shu of Fang Ji Zhuan of the literature San Guo Zhi《三國志‧魏書‧方技傳》has the record of Ma Jun (馬鈞) who built a paddle blade machine [46]: "There was a smart person called Ma Jun from Fu Feng. … The city had lands for gardening but there was no water. Ma Jun built a device that could be operated by children to draw water. The water went in at one end and comes out the other. The technology of his device was significantly more advanced than those of other mechanisms." 『時有扶風馬鈞，巧思絕世。… 居京都，城內有地，可以爲園，患無水以灌之。乃作翻車，令兒童轉之，而灌水自覆，更入更出，其巧百倍於常。』

Based on the way they were operated, the manually driven paddle blade machines were classified as hand-operated and foot-operated. Figure 3.30 shows a foot-operated paddle blade machine. The book Nong Shu has a detailed record of the machine [28]: "Fan Che was referred to as the dragon spine, which was used to irrigate farm. In addition to the railings and banisters, the device used wood planks to form a groove that could be as long as two zhang. The width varied from four to seven cun, while the height was approximately one chi. Inside the groove was a conveyor, which was as wide as the groove width. It was shorter than the groove boards by one chi and attached to the large and small wheel axles. Paddle blades were attached to the conveyor. Four pegs were attached to the horizontal beam, which comes out of both sides of the larger axle and was set on racks on the bank. When a person stepped on the pegs, the conveyor with the paddle blades moved and carried water up the bank. The device had few key elements and may be easily constructed by the carpenter. Three devices may be used for banks three zhang high. Water may be drawn from a small pool at the middle of the dragon spine series to irrigate dry farms three zhang high. The device was applicable where the land was near a water source." 『翻車，今人謂之龍骨車也。 … 今農家用之漑田。其車之制，除壓欄木及列檻樁外，車身用板作槽，長可二丈，闊則不等，或四寸至七寸，高約一尺。槽中架行道板一條，隨槽闊狹，比槽板兩頭俱短一尺，用置大小輪軸，同行道板上下通周以龍骨、板葉。其在上大軸兩端，各帶拐木四茎，置於岸上木架之間。人憑架上踏動拐木，則龍骨、板隨轉，循環行道板刮水上岸。此車關鍵頗少，必用木匠，可易成造。其起水之法，若岸高三丈有餘，可用三車，中間小池倒水上之，足救三丈以上高旱之田。凡臨水地段，皆可置用。』

The books Nong Zheng Quan Shu [20], Tian Gong Kai Wu [4], and Lu Ban Jing《魯班經》[47] contain records of the paddle blade machine. Tian Gong Kai Wu states [4]: "When water from the lake did not flow, the wheel was turned by a cow or by several persons using foot pedals. A longer device was approximately two zhang; a shorter device was half the size. Inside the device was a chain of boards that were operated to push water upwards. One person's effort in a day could irrigate approximately one-third mu (畝, ancient Chinese area of acre) of land. A cow's effort irrigated double the area."『其湖池不流水，或以牛力轉盤，或聚數人踏轉。車身長者二丈，短者半之，其內用龍骨拴串板，關水逆流而上。大抵一人竟日之力，灌田五畝，而牛則倍之。』 Figure 3.20 shows a cow-driven paddle blade machine.

Figure 3.30 A foot-operated paddle blade machine [4]

The book Tian Gong Kai Wu also has records of the wind-driven paddle blade machine [4]: "In Yang Prefecture, a wind sail was used to power the device. When the wind stopped, the device ceased to operate." 『揚郡以風帆數扇，俟風轉車，風息則止。』Based on the position of the axle, windmills in ancient China were divided into horizontal- and vertical-axle windmills. The invention of the windmill dates back to the East Han Period (AD 25–220) and the Song Dynasty (AD 960–1279). Since the capacity of the windmill was dependent on the wind velocity, the wind-driven paddle wheel machines were more often used to draw water than to irrigate.

Jing che (井車)

The water machine Jing che was used to draw water from water wells; it was also called a wooden dipper water machine. The machine used wooden dippers instead of paddle blades. A series of wooden dippers were connected by a chain, and the chain was connected to a vertical wheel installed at the mouth of the water well. When the vertical wheel rotated, the wooden dippers were continuously raised to scoop the water, achieving a conveying function. Its main difference with the paddle wheel machine is that it did not have a chain wheel at the lower portion. Figure 3.31 shows a design structure of a jing che. Since it was not possible to use wooden paddles to scour water from a vertical well, a series of wooden dippers were used instead. The dippers were connected to the large wheel on the mouth of the water well. At one end of the wheel axle was a large vertical gear, which was connected to a large horizontal gear. An animal was used to rotate the large horizontal gear, which in turn drove the large vertical gear. The large vertical gear caused the large wheel connected to the chain of dippers to move, too. In this way, the dippers were continuously raised and water was brought up, deposited in a pan inside the wheel and then channeled to the field.

The publication Tai Ping Guang Ji《太平廣記》, an excerpt from the publication Qi Yan Lu《啓顏錄》, states [48]: "Deng Xuan-ting went to burn incense in the temple. He observed a water wheel in the orchard with several monks. The wooden buckets were attached in a chain to draw water from the well." 『鄧玄挺入寺行香，與諸僧詣圓觀植蔬，見水車，以木桶相連，汲于井中。』Also, according to the publication Jiu Tang Shu《舊唐書》[49], Deng Xuan-ting (鄧玄挺) *had offended during the first year of the Yung Chang Period (689 AD), and died in prison.* 『鄧玄挺「永昌元年得罪，下獄死。」』It can therefore be inferred from these that the water machine was already being used during the early years of the Tang Dynasty (AD 618–906).

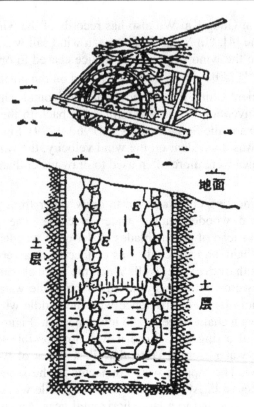

Figure 3.31 Design structure of a jing che (井車) [5]

Gao zhuan tong che (高轉筒車)

The water-lifting device named gao zhuan tong che was a chain conveyor water machine. The book Nong Shu states [28]: "The gao zhuan tong che (water-lifting device), as a standard, was ten zhang high. A higher rack and a lower rack were erected to which two wheels were installed. The lower wheel was half submerged in water. The diameter of the wheels was four chi. The two sides of the mechanism were raised high with a groove in the middle to pass the tube cable. The tube cable was arranged in three sections but connected as a ring without ends. Bamboo tubes were tied to the rope at five cun interval. Each tube was one chi long. Wood tablets of the same length were also attached below the tube cable. A metal cable was often used and wrapped around the upper and lower wheels. A flat wooden plank was attached between the wheels to absorb the weight of the tube cable. The machine was driven by foot paddles or a cow rotating the upper wheel. The tube cable scooped water from below and traveled to the upper wheel along the groove. The tubes deposited water upon reaching the upper wheel and empty tubes were returned below. This process continued, and the water drawn in a day was no less than that those drawn using a hu che on flat ground. If it were a pool, another machine would be attached, making a total length of over 200 chi. This was applicable when the banks were high or the fields were on a hill. The machine was being used in the pond inside Hu Qiu Temple in Ping Jiang, but because the water drawn was enough for drinking and not irrigating, it was discontinued. Recently, better methods are being tested, and the users relate their experience."

『高轉筒車，其高以十丈爲准，上下架木，各豎一輪，下輪半在水內。各輪徑可四尺。輪之一周，兩旁高起，其中若槽，以受筒索。其索用竹，均排三股，通穿爲一，隨車長短，如環無端。索上相離五寸，俱置竹筒。筒長一尺，筒索之底，拖以木牌，長亦如之。通用鐵線縛定，隨索列次，絡于上下二輪。復于二輪筒索之間，架刓木平底行槽一連，上與二輪相平，以承筒索之重。或人踏，或牛拽轉上輪，則筒索自下兜水循槽至上輪，輪首覆水，空筒復下。如此循環不已，日所得水，不減平地車戽。若積爲池沼，再起一車，計及二百餘尺。如因高岸深，或田在山上，皆可及之，今平江虎邱寺劍池亦類此制，但小小汲飲，不足漑田，故不錄。此近創捷法，已經較試，庶用者述之。』

It is clear from these writings that the water-lifting device gao zhuan tong che was composed of upper and lower wheels, a tube cable, and supporting frame. Half of the lower wheel was immersed in water, and the height to which the water was drawn could reach ten zhang. Based on the sources of power, the machine may be divided into the manpower-driven, animal-driven, and water-driven. Figure 3.32 shows a water-driven gao zhuan tong che.

Tian ti (sky ladder) (天梯)

There are some historical records on chain drives being used in power transmissions. For instance, the silver-operated clock constructed by Zhang Si-xun (張思訓) in AD 987 employed a chain drive to transmit power. In the case of the astronomical tower constructed by Su Song (蘇頌) and Han Gong-lien (韓公廉) in the early year of the Yuan You Period (AD 1086–1092) in the Northern Song Dynasty (AD 960–1127), since the vertical main shaft was too long, it was replaced by a ring chain for transmission to serve as the power source for driving the astronomical device. This device was called the **tian ti** (sky ladder). It was a typical chain drive with metallic chains for transmitting motion and force.

Figure 3.32 A water-driven gao zhuan tong che [4]

Figure 3.33 shows such a device in the book 《新儀象法要》 [50]. In the device, the rotation of the driving axle was transmitted to the upper horizontal axle through two small chain rings. This caused three gears to move the celestial movement hoop and the sun-moon-star panel of the machine. The article states: "The one zhang nine chi five cun long 'sky ladder' was a chain of connected metal braces. The chain was connected to the upper and lower hubs. Every turn of the chain caused the celestial movement hoop to move a distance, which in turn caused the sun-moon-star panel to move, too." 『天梯，長一丈九尺五寸。其法以鐵括聯周匝上，以鰲云中天梯上轂掛之。下貫樞軸中天梯下轂。每運一括則動天運環一距，以轉三辰儀，隨天運動。』 The term tie gua (鐵括) in the book refers to spare parts of the metallic chain. The upper and lower hubs refer to the small chain wheels in the upper and lower axles. The sky ladder and chain wheels could be used to accurately transfer force. Their effects and those found in modern machines are entirely the same. They are actual examples of the earliest application of chain drives in ancient China.

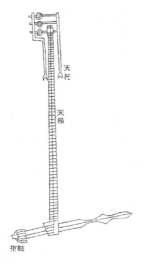

Figure 3.33 Tian ti (sky ladder) [50]

References

1. Edited by Lu, J.Y. and Hua, J.M., A History of Science and Technology in China – Volume of Mechanical Engineering (in Chinese), Science Press, Beijing, 2000.
陸敬嚴，華覺明主編，中國科學技術史・機械卷，科學出版社，北京，2000年。

2. Lu, J.Y., History of Chinese Machines (in Chinese), Ancient Chinese Machinery Cultural Foundation (Tainan, Taiwan), Yue Yin Publishing House, Taipei, 2003.
陸敬嚴，中國機械史，中華古機械文教基金會(台南，台灣)，越吟出版社，台北，2003年。

3. Kao Gong Ji (in Chinese), annotated by Zheng Xuan (Han Dynasty), commentaries by Jia Gong-yan (Tang Dynasty), collated by Ruan Yuan (Qin Dynasty), notes and commentaries from Zhou Li, Chapter 41, Da Hua Publishing House, Taipei, 1989.
《考工記》；鄭玄[漢朝]注，賈公彥[唐朝]疏，阮元[清朝]校勘，周禮注疏，卷四十一，大化出版社，台北，1989年。

4. Tian Gong Kai Wu (in Chinese) by Song Ying-xing (Ming Dynasty), Taiwan Commercial Press, Taipei, 1983.
《天工開物》；宋應星[明朝]撰，天工開物，台灣商務印書館，台北，1983年。

5. Liu, X.Z., History of Inventions in Chinese Mechanical Engineering (in Chinese), Science Press, Beijing, 1962.

　　劉仙洲，中國機械工程發明史(第一篇)，科學出版社，北京，1962年。

6. Yan, H.S., "Ancient Chinese wedges (in Chinese)," Mechanical Engineering, Chinese Society of Mechanical Engineers, Taipei, No. 221, pp. 34–37, February 1998.
　　顏鴻森，古中國的尖劈，機械工程，中國機械工程學會，台北，第 221 期，34–37頁，1998年02月。

7. Gu Jin Shi Wu Kao (in Chinese) by Wang San-pin (Ming Dynasty), Taiwan Commercial Press, Taipei, 1973.
　　《古今事物考》；王三聘[明朝]撰，台灣商務印書館，台北，1973年。

8. Gu Shi Kao (in Chinese) by Qiao Zhou (Three Kingdoms), Yi Wen Publishing House, Taipei, 1962.
　　《古史考》；譙周[蜀]撰，藝文出版社，台北，1962年。

9. Shi Wu Gan Zhu (in Chinese) by Huang Yi-zheng (Ming Dynasty), Qi Lu Books, Jinan, 1955.
　　《事務紺珠》；黃一正[明朝]撰，齊魯書社，濟南，1995年。

10. Shuo Yuan (in Chinese) by Liu Xiang (Han Dynasty), Taiwan Commercial Press, Taipei, 1965.
　　《說苑》；劉向[漢朝]撰，台灣商務印書館，台北，1965年。

11. Lun Heng (in Chinese) by Wang Chong (Han Dynasty), Hong Ye Books, Taipei, 1983.
　　《論衡》；王充[漢朝]撰，宏業書局，台北，1983年。

12. Yan, H.S., "Ancient Chinese inclined planes (in Chinese)," Mechanical Engineering, Chinese Society of Mechanical Engineers, Taipei, Nos. 222 and 223, pp. 56–60, June 1998.
　　顏鴻森，古中國的斜面，機械工程，中國機械工程學會，台北，第 222、223期，56–60頁，1998年06月。

13. Xun Zi (in Chinese) by Xun Kuang (Zhou Dynasty), Taiwan Commercial Press, 1967.
　　《荀子》；荀況[周朝]撰，台灣商務印書館，台北，1967年。

14. Chen, M.D. and others, Brief Stories of History of Science and Technology in China (in Chinese), Ming Wen Books, Taipei, 1992.
　　陳美東等，簡明中國科學技術史話，明文書局，台北，1992年。

15. Yan, H.S., "Ancient Chinese screws (in Chinese)," Mechanical Engineering, Chinese Society of Mechanical Engineers, Taipei, No. 224, pp. 30–33, August 1998.
　　顏鴻森，古中國的螺旋，機械工程，中國機械工程學會，台北，第 224 期，30–33頁，1998年08月。

16. Bao Po Zi (in Chinese) by Ge Hong (Jin Dynasty), Taiwan Commercial Press, Taipei, 1979.
　　《抱朴子》；葛洪[晉朝]撰，台灣商務印書館，台北，1979年。

17. San Cai Tu Hui (in Chinese) by Wang Qi (Ming Dynasty), Zhuang Yan Culture Co., Tainan, Taiwan, 1995.
　　《三才圖會》；王圻[明朝]撰，莊嚴文化事業公司，台南，台灣，1995年。

18. Wang, Z.D., Ke Ji Kao Gu Lun Cong (Papers in Technical Archaelogy) (in Chinese), Cultural Relics Publishing House, Beijing, 1963.
王振鐸，科技考古論叢，文物出版社，北京，1936 年。

19. Yan, H.S., "Ancient Chinese levers (in Chinese)," Mechanical Engineering, Chinese Society of Mechanical Engineers, Taipei, No. 225, pp. 68–73, October 1998.
顏鴻森，古中國的槓桿，機械工程，中國機械工程學會，台北，第 225 期，68–73 頁，1998 年 10 月。

20. Nong Zheng Quan Shu (in Chinese) by Xu Guang-qi (Ming Dynasty), Taiwan Commercial Press, Taipei, 1968.
《農政全書》；徐光啓[明朝]撰，台灣商務印書館，台北，1968 年。

21. Zhuang Zi (in Chinese) by Zhuang Zhou (Zhou Dynasty), Jin Xiu Publishing House, Taipei, 1993.
《莊子》；莊周[周朝]撰，錦繡出版社，台北，1993。

22. Qi Min Yao Shu (in Chinese) by Jia Si-xie (Late Wei Dynasty), Taiwan Commercial Press, Taipei, 1968.
《齊民要術》；賈思勰[後魏]撰，台灣商務印書館，台北，1968 年。

23. Lu Shi Chun Qiu (in Chinese) by Lü Bu-wei (Zhou Dynasty), Jin Xiu Publishing House, Taipei, 1993.
《呂氏春秋》；呂不韋[周朝]撰，錦繡出版社，台北，1993 年。

24. Mohist Canon (in Chinese) by Zhao Guan-zhi (Song Dynasty), Yi Wen Publishing House, Taipei, 1966.
《墨經》；晁貫之[宋朝]撰，藝文出版社，台北，1966 年。

25. Mencius (in Chinese) by Meng Ke (Zhou Dynasty), Yi Wen Publishing House, Taipei, 1969.
《孟子》；孟軻[周朝]撰，藝文出版社，台北，1969 年。

26. Yan, H.S., "Ancient Chinese pulleys (in Chinese)," Mechanical Engineering, Chinese Society of Mechanical Engineers, Taipei, No. 226, pp. 21–26, December 1998.
顏鴻森，古中國的滑輪，機械工程，中國機械工程學會，台北，第 226 期，21–26 頁，1998 年 12 月。

27. Wu Yuan (in Chinese) by Lou Qi (Ming Dynasty), Yi Wen Publishing House, Taipei, 1965.
《物原》；羅頎[明朝]撰，藝文出版社，台北，1965 年。

28. Nong Shu (in Chinese) by Wang Zhen (Yuan Dynasty), Taiwan Commercial Press, Taipei, 1968.
《農書》；王禎[元朝]撰，台灣商務印書館，台北，1968 年。

29. Jin Shi (in Chinese) by Pei Song-zhi (Southern and Northern Dynasty), Yi Wen Publishing House, Taipei, 1972.
《晉史》；裴松之[南北朝]撰，藝文出版社，台北，1972 年。

30. Wu Jing Zong Yao (in Chinese) by Zeng Gong-liang (Northern Song Dynasty), The Commercial Press, Shanghai, 1935.
《武經總要》；曾公亮[北宋]撰，商務印書館，上海，1935 年。

31. Wang Ning, "Investigation on the story of si shui qu ding on the stone painting of the Hang Dynasty," National Palace Museum Periodicals, Taipei, Vol. 22, No. 9, pp. 112–119, December 2004.
王寧，"九鼎‧泗水取鼎 – 漢畫像石「泗水取鼎」故事考實"，故宮文物月刊，台北，第 22 卷，第 9 期，第 112–119 頁，2004 年。

32. Yan, H.S., Mechanisms (in Chinese), 3rd edition, Dong Hua Books, Taipei, 2006.
顏鴻森，機構學，第三版，東華書局，台北，2006 年。

33. Hou Han Shu · Du Shi Zhuan (in Chinese) by Fan Ye (Eastern Jin Dynasty), Ding Wen Publishing House, Taipei, 1977.
《後漢書‧杜詩傳》；范曄[東晉]撰，鼎文出版社，台北，1977 年。

34. Huan Zi Xin Lun (in Chinese) by Huan Tan (Han Dynasty), Yi Wen Publishing House, Taipei, 1967.
《桓子新論》；桓譚[漢朝]撰，藝文出版社，台北，1967 年。

35. Jin Zhu Gong Zan (in Chinese) by Jin Fu-chang (Jin Dynasty), Yi Wen Publishing House, Taipei, 1972.
《晉諸公讚》；晉傅暢[晉朝]撰，藝文出版社，台北，1972 年。

36. Song Shi (History of the Song Dynasty) (in Chinese) by Tuo Tuo (Yuan Dynasty), Vol. 340, Ding Wen Publishing House, Taipei, 1983.
《宋史》；脫脫[元朝]等撰，卷三百四十，鼎文出版社，台北，1983 年。

37. Yuan Wen Lei (in Chinese), edited by Su Tian-jue (Yuan Dynasty), World Books, Taipei, 1962.
《元文類》；蘇天爵[元朝]編，世界書局，台北，1962 年。

38. Ming Shi (in Chinese) by Zhang Ting-yu (Qin Dynasty), Jin Xiu Publisher, Taipei, 1993.
《明史》；張廷玉[清朝]撰，錦繡出版社，台北，1993 年。

39. Xiao Er Ya (in Chinese) by Kong Fu (Han Dynasty), Yi Wen Publishing House, Taipei, 1966.
《小爾雅》；孔鮒[漢朝]撰，藝文出版社，台北，1966 年。

40. Yi-Jing (in Chinese) by unknown author, Archeology Publishing House, Taipei, 1985.
《易經》；佚名，考古出版社，台北，1985 年。

41. Shuo Wen (in Chinese) by Xu Shen (Han Dynasty), Yi Wen Publishing House, Taipei, 1959.
《說文》；許慎[漢朝]撰，藝文出版社，台北，1959 年。

42. Tian Shu (in Chinese) by Qin-Guan (Song Dynasty), Yi Wen Publishing House, Taipei, 1971.
《蠶書》；秦觀[宋朝]撰，藝文出版社，台北，1971 年。

43. Guang Tian Sang Shuo Ji Yao (in Chinese) by Shen Lian (Qin Dynasty), Yi Wen Publishing House, Taipei, 1965.
《廣蠶桑說輯要》；沈練[清朝]撰，藝文出版社，台北，1965 年。

44. General Study of Calabash Instruments in the Shang and Zhou Dynasties (in Chinese) by Rong Geng, Supplement of Yenching Journal of Chinese Studies, Vol. 17 &18, Oriental Culture, Taipei, 1973.

商周蠱器通考，容庚，燕京學報專號，第 17、18 冊，東方文化，台北，1973 年。

45. Hou Han Shu (in Chinese) by Fan Ye (Eastern Jin Dynasty), Ding Wen Publishing House, Taipei, 1977.

《後漢書》；范曄[東晉]撰，鼎文出版社，台北，1977 年。

46. San Guo Zhi (in Chinese) by Chen Shou (Jing Dynasty), Yee Wen Publisher, Taipei, 1958.

《三國志》；陳壽[晉朝]撰，藝文出版社，台北，1958 年。

47. Hui Tu Lu Ban Jing (in Chinese), edited by Wu Rong (Ming Dynasty), Zhu Lin Books, Hsinchu, 1995.

《繪圖魯班經》；午榮[明朝]彙編，竹林書局，新竹，1995 年。

48. Tai Ping Guang Ji (in Chinese), edited by Li Fang (Song Dynasty), Taiwan Commercial Press, Taipei, 1983.

《太平廣記》；李昉[宋朝]編，台灣商務印書館，台北，1983 年。

49. Jiu Tang Shu (in Chinese) by Liu Xu (Eastern Jin Dynasty), Ding Wen Publishing House, Taipei, 1976.

《舊唐書》；劉昫[東晉]撰，鼎文出版社，台北，1976 年。

50. Xin Yi Xiang Fa Yao (in Chinese) by Su Song (Northern Song Dynasty), Taiwan Commercial Press, Taipei, 1969.

《新儀象法要》；蘇頌[北宋]撰，新儀象法要，台灣商務印書館，台北，1969 年。

Chapter 4 Reconstruction Design Methodology

This chapter presents a methodology for the systematic reconstruction of design concepts of lost machines. Design specifications of the target subject are addressed. Various atlases of generalized kinematic chains are introduced and provided as the inclusive data banks for the generation of feasible specialized chains. The concept of specialization is further presented for identifying all possible reconstruction designs. An example regarding the conceptual design of planetary gear trains for infinitely various transmissions [1, 2] is adopted simply for the purpose of illustrating each step of the design methodology in detail.

4.1 Introduction

One of the most difficult tasks in research of lost ancient machines is to reconstruct the original designs.

In past years some reconstruction designs of lost machines in ancient China were brought into existence based on literature studies, and with or without the help of modern science and technology. However, these designs were mainly based on personal knowledge and judgment. The process relied on individual experiences, and the results may not be solidly functional and proven. So far, no existing method is available to directly guide researchers to reinvent lost designs. The purpose of this chapter is to present a step-by-step procedure for the systematic reconstruction of all possible topological structures of lost machines, subject to design requirements and constraints. This procedure utilizes the idea of creative mechanism design methodology to converge the divergent conceptions obtained from literature studies to a focused scope. Mechanical evolution and the variation method are then applied to obtain feasible reconstruction designs that meet the science and technology standards of the subject's time period. [2–4]

This methodology does not attempt to reconstruct the original design for a specific lost machine. Its purpose is to systematically generate all possible design concepts, that is, the topological structure of mechanisms of the target lost machine. If the defined and/or concluded design specifications, topological characteristics, and design requirements and constraints are feasible, one of the resulting reconstruction designs should be the original design. Such an approach provides a logical tool for historians in ancient machinery to further identify the possible original designs based on proven historical archives.

4.2 Procedure of Reconstruction Design

A methodology is a set of procedures or design steps for systematically solving a defined problem. It includes a body of methods, rules, and/or postulates. Figure 4.1 shows the flowchart for the reconstruction design methodology of lost machines. It includes the following four steps:

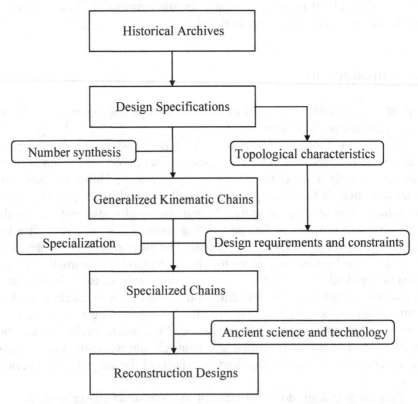

Figure 4.1 Procedure of the reconstruction design methodology

Step 1. Develop design specifications of the target lost machine based on the study of available historical archives, and conclude the topological characteristics of the design.

Step 2. Obtain the atlas of generalized kinematic chains with the required numbers of members and joints as specified in the concluded topological characteristics of the design in Step 1, based on the algorithm of number synthesis [2].

Step 3. Assign required types of members and joints to each generalized kinematic chain obtained in Step 2, and based on the process of specialization, to have the atlas of specialized chains subject to design requirements and constraints concluded from the topological characteristics of the design.

Step 4. Particularize each specialized chain obtained in Step 3 into its corresponding schematic format to have the atlas of reconstruction designs that meet the science and technology standards of the subject's time period by utilizing the mechanical evolution and variation theory to perform a mechanism equivalent transformation.

In what follows, each step of the design methodology is explained in detail.

4.3 Design Specifications

The reconstruction of ancient machines requires exhaustive literature study to clearly recognize and define the problem in order to develop design specifications. It is also important to be familiar with the available science and technology of the subject's time period. Mechanical elements and mechanisms of lost ancient machines may be different in different dynasties.

Specification is a precise written statement describing a machine, such as its topological structure, motion type and range, force scale, driving power, work capacity, efficiency, and performance. It should be defined in the beginning of the design process of any machine. Without a clear statement of specifications, the design process of the machine cannot be solidly carried out. Different machines, or machines with different performance functions, have different specifications. Any solution that does not meet the specifications is worthless.

In the early stage of the conceptual phase for the reconstruction design of lost machines, only basic specifications regarding topological structures of mechanisms are of major concern. Those items relating to the dimension and the state of motion (position, velocity, and acceleration) of members and mechanisms can be disregarded at this stage.

Based on the developed design specifications or by studying the topo-logical characteristics of mechanisms of available existing designs, design requirements and constraints should be concluded. Design requirements and constraints can be flexible and are varied for different cases and ex-pectations. They are normally identified based on technology reality and designers' decisions. Different design requirements and constraints result in different atlases of specialized chains.

[Example 4.1]
Topological characteristics of planetary gear trains for infinitely variable transmissions.

Planetary gear trains are mechanisms in which at least one member is required to rotate about its own axis and at the same time to revolve about another axis to provide constant velocity ratios. They are used in various transmission systems due to the advantages of being lightweight, compact, and having a high gear ratio. One application of planetary gear trains is in infinitely variable transmissions.

In general, an infinitely variable transmission (IVT) consists of a con-tinuous variable unit (CVU) and a planetary gear train (PGT) with two de-grees of freedom. Since an IVT has the kinematic property of zero velocity ratios, it can be used as the power train for exercising machines. Figure 4.2 shows the schematic drawing of such a design. It consists of an input-coupled CVU and a PGT with two degrees of freedom.

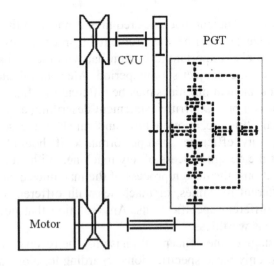

Figure 4.2 An infinitely variable transmission

An important characteristic that governs the performance of infinitely variable transmissions is the topological structure of the planetary gear train. The simplest planetary gear train with two degrees of freedom has five members. Figure 4.3 shows the schematic representation of a five-bar planetary mechanism. It has one carrier (member 2, K_{Lc}), adjacent to the ground link (member 1, K_F) with a revolute joint (joint a, J_R) and to a planet gear (member 3, K_{Gp}) with a revolute joint (joint d, J_R), as the output. It has one sun gear (member 4, K_{Gs}), adjacent to the ground link with a revolute joint (joint b, J_R) and meshing with the planet gear with a gear joint (joint e, J_G), as one input (input I). It has one ring gear (member 5, K_{Gr}), adjacent to the ground link with a revolute joint (joint c, J_R) and meshing with the planet gear with a gear joint (joint f, J_G), as another input (input II). The topology matrix (M_T) of this mechanism, as defined in Chapter 2, is:

$$M_T = \begin{bmatrix} K_F & J_R & 0 & J_R & J_R \\ a & K_{Lc} & J_R & 0 & 0 \\ 0 & d & K_{Gp} & J_G & J_G \\ b & 0 & e & K_{Gs} & 0 \\ c & 0 & f & 0 & K_{Gr} \end{bmatrix}$$

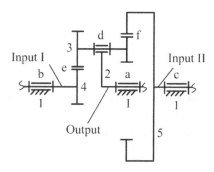

Figure 4.3 A five-bar planetary gear train (Example 4.1)

By studying this design, it is concluded that the planetary gear train for the infinitely variable transmission has the following topological characteristics:

1. It has five members, consisting of at least one ground link, one carrier, one planet gear, one sun gear, and one ring gear.
2. It has six joints including four revolute joints and two gear joints.
3. It is a reverted gear train, i.e., the input and output members are coaxial.
4. The sun gear and the ring gear are the two inputs, and the carrier is the output.

5. All gears are spur gears.

6. It has two degrees of freedom.

Based on Equation (2.1) for mobility analysis, the following expression is true for a planetary gear train with two degrees of freedom ($F_p = 2$), N_L members, N_{JR} revolute joints, and N_{JG} gear joints:

$$2N_{JR} + N_{JG} - 3N_L + 5 = 0 \qquad (4.1)$$

It is obvious that the following expression is also true for a planetary gear train with N_J joints:

$$N_{JR} + N_{JG} - N_J = 0 \qquad (4.2)$$

Furthermore, every link in a geared kinematic chain has at least one revolute pair. By removing all gear joints from the geared chain, it forms a tree-chain, i.e., a connected chain without any loop. Based on the fact that a tree-chain with N_L links contains $N_J - 1$ joints from graph theory, the following expression for a planetary gear train is true [5]:

$$N_{JR} - N_L + 1 = 0 \qquad (4.3)$$

By solving Equations (4.1)–(4.3) for the numbers of joints (N_J, N_{JR}, and N_{JG}) in terms of the number of links (N_L), the following relations for planetary gear trains with two degrees of freedom are concluded:

$$N_J = 2N_L - 4 \qquad (4.4)$$

$$N_{JR} = N_L - 1 \qquad (4.5)$$

$$N_{JG} = N_L - 3 \qquad (4.6)$$

Equations (4.4)–(4.6) indicate that for a planetary gear train with two degrees of freedom and with five members, it always has six joints, consisting of four revolute joints and two gear joints.

4.4 Generalized Kinematic Chains

The second step of the reconstruction design methodology is to obtain the atlas of generalized kinematic chains with the required numbers of members and joints as specified in the concluded topological characteristics of mechanisms.

A generalized joint is a joint in general; it can be a revolute joint, prismatic joint, spherical joint, helical joint, or some others [2]. A generalized joint with two incident members is called a simple generalized joint, whereas a generalized joint with more than two incident members is called a multiple generalized joint. A generalized joint with N_L incident members is graphically symbolized by $N_L - 1$ small concentric circles. Figure 4.4(a) and (b) show a generalized joint with two and three incident members, respectively. In addition, Figure 4.4(a) is a simple generalized joint, and Figure 4.4(b) is a multiple generalized joint.

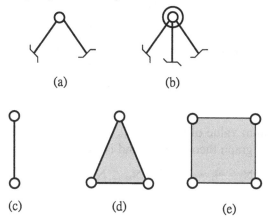

(a)   (b)

(c)   (d)   (e)

Figure 4.4 Graphical representations of generalized joints and links

A generalized link is a link with generalized joints; it can be a binary link, ternary link, quaternary link, etc. Graphically, a generalized link with N_J incident joints is symbolized by a shaded N_J-sided polygon with small circles as vertices. Figure 4.4(c)–(e) show a binary, ternary, and quaternary generalized link, respectively.

A generalized kinematic chain consists of generalized links connected by generalized joints. It is connected, closed, without any bridge-link, and with simple joints only [2]. An (N_L, N_J)-generalized kinematic chain refers to a generalized kinematic chain with N_L-generalized links and N_J generalized joints. The topological structure of a generalized kinematic chain is characterized by the number and the type of links, the number of joints, and the incidences between links and joints; and can be represented by its topology matrix, M_T, as defined in Session 2.6. Furthermore, if all joints of a generalized kinematic chain are revolute joints, it becomes a kinematic chain as defined in Session 2.4.

The link assortment, A_L, of a generalized kinematic chain is the number and the type of links in the chain. It is a set of numbers consisting of the numbers of binary links (N_{L2}), ternary links (N_{L3}), quaternary links (N_{L4}), etc., and is expressed as:

$$A_L = [N_{L2}/N_{L3}/N_{L4}/...]$$

Since a generalized kinematic chain must be connected, closed, and without any bridge-link, the link assortments of a generalized kinematic chain with N_L links and N_J joints can be obtained by solving the following two equations:

$$N_{L2} + N_{L3} + \ldots + N_{Li} + \ldots + N_{Lm} = N_L \qquad (4.7)$$

$$2N_{L2} + 3N_{L3} + \ldots + iN_{Li} + \ldots + mN_{Lm} = 2N_J \qquad (4.8)$$

where N_{Li} is the number of links with i incident joints and m is the maximum number of joints incident to a link. Furthermore, the number of joints N_J is constrained by the following expression:

$$N_L \leq N_J \leq N_L(N_L - 1)/2 \qquad (4.9)$$

The maximum value of m, i.e., m_{max}, can be derived based on elementary concepts of graph theory [6, 7], and is expressed as:

$$m_{max} = \begin{cases} N_J - N_L + 2 & \text{for } N_L \leq N_J \leq 2N_L - 3 \qquad (4.10) \\ N_L - 1 & \text{for } 2N_L - 3 \leq N_J \leq N_L(N_L - 1)/2 \end{cases}$$

Based on Equations (4.7)–(4.10), all possible link assortments of generalized kinematic chains can be obtained.

[Example 4.2]
List link assortments of (5, 6) generalized kinematic chains.

For (5, 6) generalized kinematic chains, $N_L = 5$, $N_J = 6$, based on Equation (4.10), m_{max} is:

$$\begin{aligned} m_{max} &= N_J - N_L + 2 \\ &= 6 - 5 + 2 \\ &= 3 \end{aligned}$$

Therefore, Equations (4.7) and (4.8) become:

$$N_{L2} + N_{L3} = 5$$

$$2N_{L2} + 3N_{L3} = 12$$

By solving these two equations, $N_{L2} = 3$ and $N_{L3} = 2$, i.e., the corresponding link assortment is:

$$A_L = [3/2]$$

as shown in Figure 4.5(a).

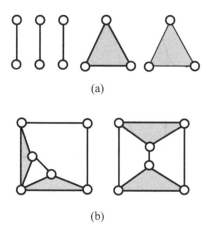

(a)

(b)

Figure 4.5 Link assortments and atlas of (5, 6) generalized kinematic chains with $A_L = [3/2]$ (Example 4.2)

[Example 4.3]
List link assortments of (6, 8) generalized kinematic chains.
  For (6, 8) generalized kinematic chains, $N_L = 6$, $N_J = 8$, based on Equation (4.10), m_{max} is:

$$m_{max} = N_J - N_L + 2$$
$$= 8 - 6 + 2$$
$$= 4$$

Therefore, Equations (4.7) and (4.8) become:

$$N_{L2} + N_{L3} + N_{L4} = 6$$

$$2N_{L2} + 3N_{L3} + 4N_{L4} = 16$$

By solving these two equations, the following three sets of solutions are available:
1. $N_{L2} = 2$, $N_{L3} = 4$, and $N_{L4} = 0$, the corresponding link assortment is $A_L = [2/4/0]$ as shown in Figure 4.6(a).
2. $N_{L2} = 3$, $N_{L3} = 2$, and $N_{L4} = 1$, the corresponding link assortment is $A_L = [3/2/1]$ as shown in Figure 4.6(b).
3. $N_{L2} = 4$, $N_{L3} = 0$, and $N_{L4} = 2$, the corresponding link assortment is $A_L = [4/0/2]$ as shown in Figure 4.6(c).
  The atlas of generalized kinematic chains with N_L links and N_J joints can be obtained by assembling link assortments with N_L links and N_J joints subject to the following constraints:

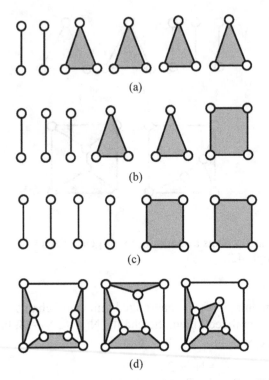

Figure 4.6 Link assortments and partial atlas of (6, 8) generalized kinematic chains (Example 4.3)

1. All links must be used to make the chain connected.
2. All joints must be used to make the chain closed.
3. No bridge-link should be formed.
4. A joint can only have two incident links to make the chain with simple joints only.
5. No link shall be adjacent to another by more than one joint.

For the link assortment of A_L = [3/2] shown in Figure 4.5(a), two (5, 6) generalized chains can be assembled as shown in Figure 4.5(b). For the link assortment of A_L = [3/2/1] shown in Figure 4.6(b), three (6, 8) generalized chains can be assembled as shown in Figure 4.6(d).

The atlas of various (N_L, N_J) generalized kinematic chains can be synthesized by assembling the corresponding link assortments. Figures 4.7–4.17 show some important atlases of generalized kinematic chains that should cover most applications for the reconstruction design of lost ancient machines.

Figure 4.7 Atlas of (3, 3) generalized kinematic chains

Figure 4.8 Atlas of (4, 4) generalized kinematic chains

Figure 4.9 Atlas of (4, 5) generalized kinematic chains

Figure 4.10 Atlas of (5, 5) generalized kinematic chains

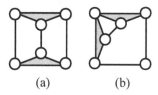

(a)   (b)

Figure 4.11 Atlas of (5, 6) generalized kinematic chains

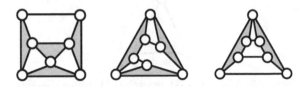

Figure 4.12 Atlas of (5, 7) generalized kinematic chains

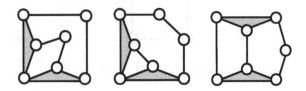

Figure 4.13 Atlas of (6, 7) generalized kinematic chains

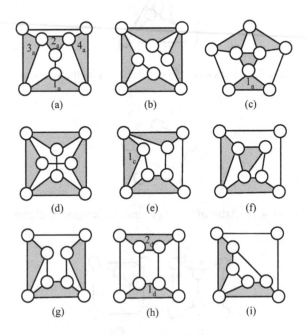

Figure 4.14 Atlas of (6, 8) generalized kinematic chains

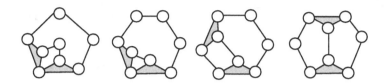

Figure 4.15 Atlas of (7, 8) generalized kinematic chains

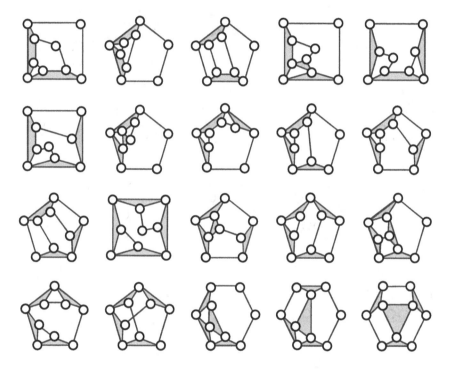

Figure 4.16 Atlas of (7, 9) generalized kinematic chains

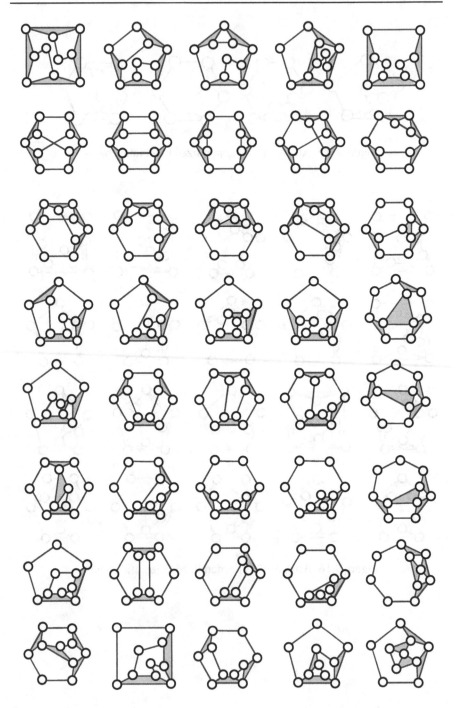

Figure 4.17 Atlas of (8, 10) generalized kinematic chains

[Example 4.4]
Generalized kinematic chains of planetary gear trains for infinitely variable transmissions.

Once the topological characteristics of the planetary gear train for infinitely variable transmissions as described in Example 4.1 are obtained, the next step in the procedure of the reconstruction design methodology is to obtain the atlas of generalized kinematic chains with five members and six joints. And, there are two (5, 6) generalized kinematic chains as shown in Figure 4.11.

Since the two (5, 6) generalized kinematic chains shown in Figure 4.11 might have limited room for generating numerous potential design concepts, planetary gear trains with two degrees of freedom and with six instead of five members for infinitely variable transmissions can be considered. In such a case, that is, for a planetary gear train with two degrees of freedom and six members, based on Equations (4.4)–(4.6), it always has eight joints: five revolute joints and three gear joints. There are also nine (6, 8) generalized kinematic chains available for the process of specialization, as shown in Figure 4.14.

4.5 Specialized Chains

The core concept of the reconstruction design methodology is specialization. The process of assigning specific types of members and joints in the available atlas of generalized kinematic chains, subject to certain design requirements and constraints, is called specialization. And, a generalized kinematic chain after specialization is called a specialized chain.

[Example 4.5]
Specialization of the (3, 3) generalized kinematic chain shown in Figure 4.18(a).

Link 1 is grounded to transform the (3, 3) generalized kinematic chain into the corresponding generalized mechanism. If joints a and b are revolute joints and joint c is a cam joint, as shown in Figure 4.18(b), based on Equation (2.1), $N_L = 3$, $C_{pR} = 2$, $N_{JR} = 2$, $C_{pA} = 1$, $N_{JA} = 1$, the degrees of freedom F_p of this mechanism is:

$$F_p = 3(N_L - 1) - (N_{JR}C_{pR} + N_{JA}C_{pA})$$
$$= (3)(3-1) - [(2)(2) + (1)(1)]$$
$$= 1$$

It is a simple cam mechanism, as shown in Figure 3.12.

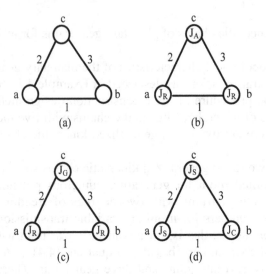

Figure 4.18 Specialization of (3, 3) generalized kinematic chain (Example 4.5)

If joints a and b are revolute joints and joint c is a gear joint, as shown in Figure 4.18(c), based on Equation (2.1), $N_L = 3$, $C_{pR} = 2$, $N_{JR} = 2$, $C_{pG} = 1$, $N_{JG} = 1$, the degrees of freedom F_p of this mechanism is:

$$F_p = 3(N_L - 1) - (N_{JR}C_{pR} + N_{JG}C_{pG})$$
$$= (3)(3 - 1) - [(2)(2) + (1)(1)]$$
$$= 1$$

It is a simple gear mechanism, as shown in Figure 3.15.

If joints a and c are spherical joints and joint b is a cylindrical joint, as shown in Figure 4.18(d), based on Equation (2.2), $N_L = 3$, $C_{sS} = 3$, $N_{JS} = 2$, $C_{sC} = 4$, $N_{JC} = 1$, the degrees of freedom F_s of this mechanism is:

$$F_s = 6(N_L - 1) - (N_{JS}C_{sS} + N_{JC}C_{sC})$$
$$= (6)(3 - 1) - [(2)(3) + (1)(4)]$$
$$= 2$$

It is specialized into a (3, 3) spatial mechanism, and it is the mechanism of the cotton thread rack shown in Figure 3.9.

[Example 4.6]
Specialization of the (4, 4) generalized kinematic chain shown in Figure 4.19(a).

Link 1 is grounded to transform the (4, 4) generalized kinematic chain into the corresponding generalized mechanism. If all joints are revolute joints, as shown in Figure 4.19(b), based on Equation (2.1), $N_L = 4$, $C_{pR} = 2$,

$N_{JR} = 4$, the degrees of freedom F_p of this mechanism is:

$$F_p = 3(N_L - 1) - (N_{JR}C_{pR})$$
$$= (3)(4 - 1) - [(2)(4)]$$
$$= 1$$

It is a simple four-bar linkage, and it is the mechanism of jie chi shown in Figures 2.1(b) and 3.11.

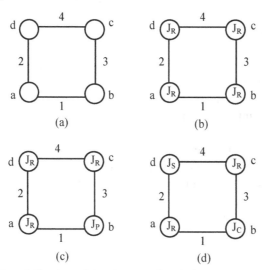

Figure 4.19 Specialization of (4, 4) generalized kinematic chain (Example 4.6)

If joints a, c, and d are revolute joints, and joint b is a prismatic joint, as shown in Figure 4.19(c), based on Equation (2.1), $N_L = 3$, $C_{pR} = 2$, $N_{JR} = 3$, $C_{pP} = 2$, $N_{JP} = 1$, the degrees of freedom F_p of this mechanism is:

$$F_p = 3(N_L - 1) - (N_{JR}C_{pR} + N_{JP}C_{pP})$$
$$= (3)(3 - 1) - [(3)(2) + (1)(2)]$$
$$= 1$$

It is a slider crank mechanism, and it is also the mechanism for traditional internal combustion engines.

If joints a and c are revolute joints, joint d is a spherical joint, and joint b is a cylindrical joint, as shown in Figure 4.19(d), based on Equation (2.2), $N_L = 4$, $C_{sR} = 5$, $N_{JR} = 2$, $C_{sS} = 3$, $N_{JS} = 1$, $C_{sC} = 4$, $N_{JC} = 1$, the degrees of freedom F_s of this mechanism is:

$$F_s = 6(N_L - 1) - (N_{JR}C_{pR} + N_{JS}C_{sS} + N_{JC}C_{sC})$$
$$= (6)(4 - 1) - [(2)(5) + (1)(3) + (1)(4)]$$
$$= 1$$

It is a spatial four-bar mechanism.

[Example 4.7]
Specialized chains of planetary gear trains for infinitely variable transmissions.

Based on the concluded topological characteristics of the planetary gear trains of infinitely variable transmissions described in Example 4.1, design requirements and constraints are concluded as follows:

Ground link
1. One of the links in each generalized kinematic chain must be the ground link.
2. A ground link must be a multiple link in order to have two input members and one output member.
3. Since a planetary transmission is a reverted gear train, a ground link must not be included in a three-bar loop.

Planet gear
1. There must be at least one planet gear.
2. Any link that is not adjacent to the ground link is a planet gear.
3. A planet gear that is not adjacent to another planet gear must not be included in a three-bar loop.
4. A planet gear must be a multiple link including at least one gear joint and one revolute joint incident to the carrier, in order to avoid degeneration.

Carrier
1. There must be a carrier corresponding to each planet gear.
2. A carrier must be adjacent to both the planet gear and the ground link.
3. Two or more planet gears in series must share a common carrier, in order to maintain the center distance between them.

Sun gear
1. There must be at least one sun gear.
2. Any link that is adjacent to the ground link and that is not a carrier is a sun gear.

Revolute joint
1. There must be $N_L - 1$ revolute joints.
2. Every link must have at least one revolute joint.
3. Any joint incident to the ground link must be a revolute joint.
4. The common incident joint of a planet gear and a carrier must be a revolute joint.
5. A planet gear can have only one revolute joint.
6. There can be no loop formed exclusively by revolute joints.

Gear joint
1. There must be $N_L - 3$ gear joints.
2. Any joint incident to both a planet gear and a sun gear must be a gear joint.
3. There can be no three-bar loop formed exclusively by gear joints.

The next step in the procedure of the reconstruction design methodology, as described in the following substeps, is to identify the corresponding specialized chains from the available atlas of generalized kinematic chains, subject to the concluded design requirements and constraints:
1. For each generalized kinematic chain, identify the ground link for all possible cases.
2. For each case obtained in substep 1, identify the planet gear(s).
3. For each case obtained in substep 2, identify the corresponding carrier(s) for all possible cases.
4. For each case obtained in substep 3, identify the sun gear(s).
5. For each case obtained in substep 4, identify the gear joints.
6. For each case obtained in substep 5, identify the revolute joints.

In what follows, the specialization for planetary gear trains with five and six members is carried out.

I. Planetary gear trains with five members
For the two (5, 6) generalized kinematic chains shown in Figure 4.11, only the one shown in Figure 4.11(a) is qualified to have a ground link, due to the constraints that a ground link should be a multiple link and should not be included in a three-bar loop. Its corresponding specialized planetary gear train can be identified as follows:
1. Since ternary links 1 and 3 are symmetric, only one of them is taken as the ground link.
2. If ternary link 1 is taken as the ground link, joints *a*, *b*, and *c* are revolute joints and ternary link 3 is the planet gear.
3. Since binary links 2, 4, and 5 are symmetric and adjacent to the planet gear, only one of them is taken as the carrier.
4. If binary link 2 is taken as the carrier, joint *d* is a revolute joint and binary links 4 and 5 are the sun gears.
5. Since there must be four revolute joints and two gear joints, the remaining two joints *e* and *f* are gear joints.

Therefore, only one specialized five-bar planetary gear train is available, as shown in Figure 4.20.

II. Planetary gear trains with six members
For the nine (6, 8) generalized kinematic chains shown in Figure 4.14, specialized chains are identified as follows:

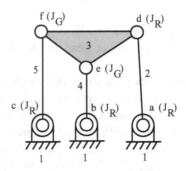

Figure 4.20 Specialized five-bar planetary gear train (Example 4.7)

Ground link

Since a ground link must be a multiple link that is not included in a three-bar loop in the chain, only generalized chains shown in Figure 4.14(a), (c), (e), and (h) are qualified to have the ground link. The ground link can be identified as follows:

1. For the generalized chain shown in Figure 4.14(a), since ternary links 1_a, 2_a, 3_a, and 4_a are symmetrical, any one of them can be taken as the ground link. Figure 4.21(a) shows its corresponding specialized mechanism with link 1_a as the ground link. In this case, joints a, b, and c are revolute joints.

2. For the generalized chain shown in Figure 4.14(c), only ternary link 1_b can be taken as the ground link. Figure 4.21(b) shows its corresponding specialized mechanism. In this case, joints a, b, and c are revolute joints.

3. For the generalized chain shown in Figure 4.14(e), only ternary link 1_c can be taken as the ground link. Figure 4.21(c) shows its corresponding specialized mechanism. In this case, joints a, b, and c are revolute joints.

4. For the generalized chain shown in Figure 4.14(h), since quaternary link 1_d or 2_d are symmetric, either one is taken as the ground link. Figure 4.21(d) shows its corresponding specialized mechanism with link 1_d as the ground link. In this case, joints a, b, c, and d are revolute joints.

Therefore, four specialized chains with one identified ground link are available as shown in Figure 4.21(a)–(d).

Planet gear

Based on the four specialized mechanisms with the ground link identified shown in Figure 4.21(a)–(d), planet gears can be identified as follows:

1. For the one shown in Figure 4.21(a), since link 2 is not a multiple link, this case is irrelevant.

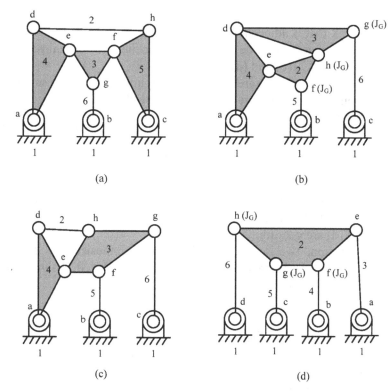

Figure 4.21 Specialized six-bar planetary gear trains (Example 4.7)

2. For the one shown in Figure 4.21(b), since ternary links 2 and 3 are not adjacent to the ground link and are multiple links, they are planet gears.
3. For the one shown in Figure 4.21(c), since link 2 is not a multiple link, this case is irrelevant.
4. For the one shown in Figure 4.21(d), since quaternary link 2 is not adjacent to the ground and is a multiple link, it is a planet gear.

Carrier

Since there must be a carrier corresponding to each planet gear and two planet gears in series must share a common carrier, carriers can be identified from the two specialized chains shown in Figure 4.21(b) and (d) as follows:

1. For the one shown in Figure 4.21(b), ternary link 4 is the common carrier to planet gears 2 and 3, and joints *d* and *e* are revolute joints.
2. For the one shown in Figure 4.21(d), binary link 3 is the carrier to planet gear 2, and joint *e* is a revolute joint.

Sun gear
Since any link that is adjacent to the ground link and that is not a carrier is a sun gear, sun gears can be identified from the two specialized chains shown in Figure 4.21(b) and (d) as follows:
1. For the one shown in Figure 4.21(b), binary links 5 and 6 are sun gears.
2. For the one shown in Figure 4.21(d), binary links 4–6 are sun gears.
Revolute joint
Since there must be five revolute joints in each design, all revolute joints for the two specialized chains shown in Figure 4.21(b) and (d) are already identified.
Gear joint
Since there must be three gear joints in each design, the remaining three unassigned joints *f*, *g*, and *h* in the two specialized chains shown in Figure 4.21(b) and (d) are gear joints.

In conclusion, two specialized chains are synthesized, as shown in Figure 4.21(b) and (d), thereby satisfying the concluded design requirements and constraints.

4.6 Reconstruction Designs

Once a specialized chain is obtained, it is particularized into its corresponding design in a schematic format according to the motion and function requirements of ancient machines. In this step, the mechanical evolution and variation theory are utilized to perform a mechanism equivalent transform of the atlas of designs.

[Example 4.8]
Design concepts of planetary gear trains for infinitely variable transmissions.

Here, each feasible specialized chain of the planetary gear trains of infinitely variable transmissions obtained in Example 4.7 is particularized to obtain the corresponding schematic diagram of the planetary gear trains.

During the process of particularization, the two gears adjacent to a gear joint can be external or internal. By enumerating all possible variations of external and internal gears, numerous planetary gear trains can be obtained. For practical applications, it is very unlikely that a planet gear is in an internal form. With this additional constraint, Figure 4.22 shows the atlas of five-bar planetary gear trains for the specialized planetary gear train shown in Figure 4.22, and Figure 4.23 shows the atlas of six-bar planetary gear trains for the two feasible specialized planetary gear trains shown in Figure 4.21(b) and (d).

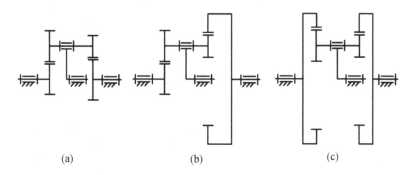

Figure 4.22 Atlas of five-bar planetary gear trains for infinitely variable transmissions (Example 4.8)

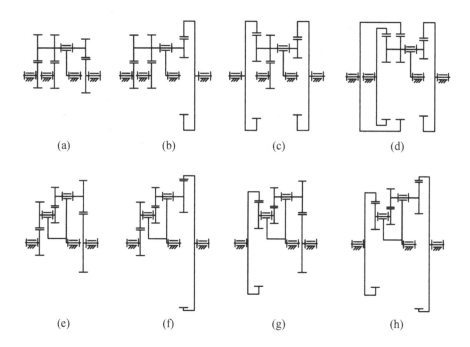

Figure 4.23 Atlas of six-bar planetary gear trains for infinitely variable transmissions (Example 4.8)

The purpose of applying ancient science theories and technologies is to identify appropriate and feasible mechanisms that can be considered as reconstruction designs. This will be addressed in detail in Chapters 5–8 for various cases for the reconstruction design of Zhang Heng's (張衡) seismoscope, Su Song's (蘇頌) escapement regulator, south-pointing chariots, and Lu Ban's (魯班) walking machines.

References

1. Yan, H.S. and Hsieh, L.C., "Concept design of planetary gear trains for infinitely variable transmissions," Proceedings of 1989 International Conference on Engineering Design, Harrogate, UK, pp. 757–766, 22–25 August 1989.
2. Yan, H.S., Creative Design of Mechanical Devices, Springer, Singapore, 1998.
3. Yan, H.S. and Lin, T.Y., "A systematic approach to the reconstruction of ancient Chinese escapement regulators," Proceedings of ASME 2002 Design Engineering Technical Conferences – the 27th Biennial Mechanisms and Robotics Conferences, Montreal, Canada, 29 September–2 October 2002.
4. Yan, H.S. and Chen, C.W., "A systematic approach for the structural synthesis of differential-type south point chariots," JSME International Journal, Series C, Vol. 49, No. 3, pp. 920–929, September 2006.
5. Buchsbaum, F. and Freudenstein, F., "Synthesis of kinematic structure of geared kinematic chains and other mechanisms," Journal of Mechanisms, Vol. 5, pp. 357–392, 1970.
6. Harary, F., Graph Theory, Addison-Wesley, Reading, MA, 1969.
7. Yan, H.S. and Harary, F., "On the maximum value of the maximum degree of kinematic chains," ASME Transactions, Journal of Mechanisms, Transmissions, and Automation in Design, Vol. 109, No. 4, pp. 487–490, 1987.

Chapter 5 Zhang Heng's Seismoscope

In AD 132, Zhang Heng (張衡) of the Eastern Han Dynasty invented the earliest seismoscope, named Hou Feng Di Dong Yi (候風地動儀). This device was recorded in literature, but lacked surviving hardware. However, several of its reconstructed designs existed in the past century.

This chapter systematically reconstructs all feasible design concepts of Zhang Heng's seismoscope that meet the scientific and technological standards of its time period. The development of ancient seismometers and western seismoscopes is introduced first. Then, historical records of Zhang Heng's seismoscope are studied and design specifications for the reconstruction design are concluded. Finally, three examples based on different design requirements and constraints are illustrated [1, 2].

5.1 Ancient Seismometers

An instrument that produces a chronologic record of ground motion during an earthquake is called a seismograph.

Seismic waves are generated by earthquakes. When an earthquake occurs, seismic waves spread out and travel in specific directions, much like waves in a pond spreading out from the point where a stone is tossed into the water. Seismographs record ground motion that results from seismic waves. The resulting seismograms provide information about the earthquake process itself and about the earth materials through which the seismic waves pass [3–5]. In general, a seismograph consists of three basic components: a seismometer, a timing system, and a recording system. The greatest challenge in the development of a seismograph was in designing a seismometer that could achieve the goal of sensing ground motion but not move with the ground. Standard practice in earthquake observatories normally use two horizontal seismometers, one oriented north–south and the other east–west, as well as a vertical seismometer. There were four types of sensing element designs in ancient horizontal seismometers, such as the common, Milne, Wiechert, and Galitzin pendulums, as shown in Figure 5.1 [1].

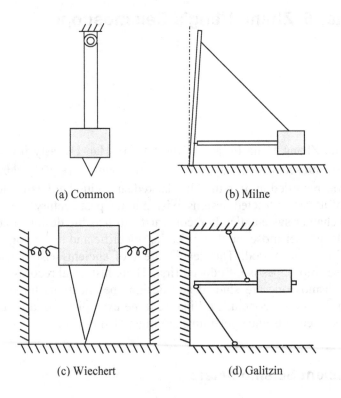

(a) Common   (b) Milne

(c) Wiechert   (d) Galitzin

Figure 5.1 Types of sensing element of ancient horizontal seismometers [1]

A common pendulum has a number of desirable characteristics as a potential seismometer. Two design principles make this a workable seismometer. First, the common pendulum is largely isolated from the ground motion by its suspension design. Second, the mass of the pendulum has inertia, and tends to remain at rest. These two basic principles were utilized in the design of early seismometers. However, the relative motion caused by a distant earthquake would be very small, and various modifications have been made to increase its sensitivity. Earlier experiments were made with longer pendulums, but later it was realized that more convenient methods of overcoming the difficulty of sensitivity are using horizontal-pendulums and extra mass. The basic principle of the horizontal pendulum is similar to a swinging gate of a fence. If the gate post is tiled off vertical, the gate describes an arc that not only has a horizontal component of movement, but also a vertical one. This arc has a very slight curvature, suggesting that it is equivalent to a vertical suspension pendulum with a very long rod length, and thus a very long period. The long period of a vertical suspension pendulum can be achieved with a horizontal suspension design.

The extra mass is to maintain pendulum stability and to return pendulum to its original position after being displaced. When the support follows earth movement, the seismic waves set up a relative movement between the frame and the pendulum. A horizontal pendulum is shown in Figure 5.1(b); it is suspended by a single wire to the arm carrying the extra mass, and the end of this arm nearer the axis has a pivot bearing. An inverted pendulum is shown in Figure 5.1(c). The extra mass carried by an upright rod is hinged to the base through two flat springs at right angles to each other, and is free to move in any horizontal direction. The instability of this arrangement can be overcome by connecting the extra mass to the frame by small springs. In the Galitzin type, Figure 5.1(d), the rod carrying the extra mass is supported from the frame by two wires. The wires are arranged such that both are pulled taut by the extra mass.

During the 19th century, many attempts were made to design an instrument for measuring the vertical movement of the ground. In some of these the mass was carried at the end of a strong horizontal spring projecting from a wall, in some it was suspended from a coiled spring, and in others it floated on a vessel of liquid. None of these instruments were satisfactory on account of the difficulty in obtaining a sufficiently long free period. As shown in Figure 5.2 [1], the problem was solved in AD 1880 by Thomas Gray who constructed an instrument in which the extra mass was fixed to one end of a lever with the spring attached between the fulcrum and the extra mass. To increase the stability of the instrument, Gray attached to the outer end of the bar a hermetically sealed tube containing mercury, which when the bar was depressed, ran outwards and increased the load in such a manner as to compensate for the decreased leverage. The combined use of horizontal and vertical seismometers resulted in a complete picture of ground motion directions.

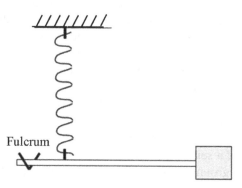

Figure 5.2 Sensing element of Gray's vertical seismometer [1]

One negative characteristic of a pendulum system used as an earthquake sensor is resonance. For a seismometer, the wildly swinging pendulum means that only the resonant vibration of the system itself is recorded, rather than the true motion of the ground. To eliminate the resonance, it is possible to control the pendulum swing through a process called damping. Early seismometers had a paddle fastened to the pendulum. The paddle was immersed in a container of a viscous fluid so that when the pendulum moved, the drag of the paddle through the fluid would mechanically damp the system, thereby eliminating wild swinging of the pendulum.

5.2 Development of Western Seismoscopes

Early earthquake instruments called seismoscopes were primarily intended to indicate that an earthquake had happened.

A seismoscope designed by J. de la Haute Feuille in AD 1703 contained a central reservoir from which mercury would spill into cups around the periphery if the ground moved [6], Figure 5.3. Andrea Bina in AD 1751 proposed a common or clock-type pendulum which was suspended over a tray of sand so that the pendulum bob could trace a record of ground motion in the sand [7], Figure 5.4. A more accurate seismoscope was built in AD 1855 by Luigi Palmieri in Italy. An electrical circuit was closed when the mercury moved, stopping a clock to indicate the time of the earthquake. A pencil pressed on a rotating drum whenever the electrical circuit was complete, providing a measure of the duration of the shaking, Figure 5.5 [8].

The first seismograph was built by Filippo Cecchi in Italy in AD 1875 [6]. Cecchi's seismograph used two common pendulums to measure horizontal motions, one swinging north–south and the other east–west, an orientation still almost always used today. The pendulum motions were magnified three times before being recorded by a rope-and-pulley mechanism, Figure 5.6. Cecchi also recorded the vertical component of motion by using a mass suspended by a spiral spring. In addition, he arranged a seismoscope to start a clock and start into motion the recording surface at the time of an earthquake. The magnification of Cecchi's seismograph was too small to record any but the strongest shakes. However, the three essential features of all useful seismographs were incorporated into this instrument. It produced a seismogram whose trace deflection was proportional to the amplitude of ground motion. The motions were amplified so that small movements could be studied, and the exact time of the event could be recorded.

Figure 5.3 J. de la Haute Feuille's seismoscope [6]

Figure 5.4 Andrea Bina's seismoscope [7]

Figure 5.5 Luigi Palmieri's seismoscope [8]

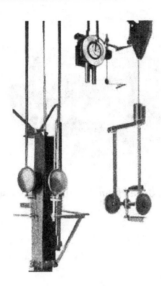

Figure 5.6 Filippo Cecchi's seismograph [6]

The most significant person in developing a practical seismograph was John Milne. In AD 1876, he joined the faculty of the Imperial College of Engineering in Tokyo. There, in association with Thomas Gray and J. Alfred Ewing, they experimented with a variety of pendulum instruments for recording ground motions. The British in Japan made many observations with their instruments and were credited with first demonstrating the value of seismographic devices to seismology. John Milne in AD 1894 used the horizontal pendulum as a sensing link, Figure 5.7 [9]. Instead of having light reflected onto photographic paper with a mirror fastened to the pendulum, Milne had light shine onto the paper through the intersection of two mutually perpendicular slits. One of the slits was fastened to the pier. The other slit was fastened to the pendulum and moved with the pendulum, thus causing the spot of light to move on the paper. Emil Wiechert used the inverted pendulum with a viscous damping as a sensing link in AD 1898 [9]. This was an important advance in seismograph design because the accuracy of the seismograph was greatly improved by the viscous damping. Moreover, because with a pendulum weight of over a ton, Wiechert's instrument had a large inertia mass and was affected very little by the friction of the stylus on the recording paper. Furthermore, the pendulum was connected to two thrust arms set at right angles with each thrust arm connected to a lever which carried a stylus for recording, Figure 5.8. Therefore, Wiechert's seismograph recorded a wide spectrum of seismic signals, accurately reproducing ground motions.

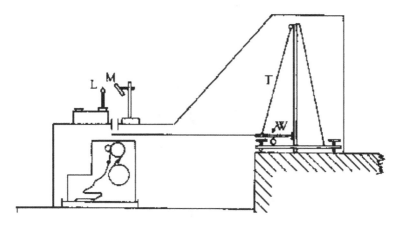

Figure 5.7 John Milne's seismograph [9]

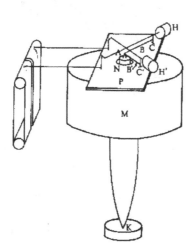

Figure 5.8 Emil Wiechert's seismograph [9]

In AD 1906 Boris Galitzin introduced a new method based on the principle that when a coil of wire moves across a magnetic field, electric currents were set up in the wire [10]. Several coils, joined in series and connected to a sensitive galvanometer, were carried by the pendulum and moved between the poles of strong magnets. As the pendulum moved the current flowing through the galvanometer was proportional to the angular velocity of the pendulum. The deflections of the galvanometer mirror were recorded photographically, Figure 5.9.

Figure 5.9 Boris Galitzin's seismograph [10]

Application of the electromagnetic principle remains the basis for most modern seismograph systems. An important addition to this is the development of digital recording in recent years. Digital recording is achieved by converting voltages into discrete numbers that represent motion by periodically sampling and recording ground motions as numbers. This can be done automatically and has the advantage of having the data in a much more useful format.

Different methods have been used to record earthquake ground motion, such as mechanical, direct optical, and galvanometric registrations. The pendulum of a seismograph which records mechanically is connected by levers or by a long arm to a scribing point which rests on a smoked sheet of paper. The mechanical registration has one great advantage, in that the records are visible while the seismograph is in operation. However, there is an uncertain amount of friction at the pivots of the magnifying levers, and between the scriber and the sheet. Thus, the primary drawback of mechanical registration is its lack of sensitivity. In direct optical registration, a beam of light is reflected from a mirror that is usually connected to the pendulum, and focused on a moving sheet of photographic paper. The light beam serves as a long weightless pointer, thereby producing no friction on the paper surface or at pivots in the magnifying levers. Therefore, the moment of inertia of the moving system is less than devices with mechanical elements. Nevertheless, the higher cost of the photographic paper was a disadvantage in the late 19th century. However, the advantages of the galvanometric registration are that the instrument is not affected by tilting, it gives a very high magnification, it achieves a great sensitivity, and it is possible to have the pendulum system and the recording device in different rooms. With the application of galvanometric registration, it is possible to detect the worldwide earthquakes occurring each year.

5.3 Zhang Heng the Man

Zhang Heng (張衡) (AD 78–139) was an extraordinary polymath in the Eastern Han Dynasty (AD 25–220) [11]. He was not only a royal astronomer but also a distinguished cartographer, mathematician, poet, painter, and inventor. He designed and built many efficient and effective instruments of which his seismoscope was the most famous.

Zhang Heng was born in the 3rd year (AD 78) of the reign of Emperor Zhang of the Eastern Han Dynasty in ancient China in county Xi-E of Nanyang. He died at the age of 62 during the 4th year (AD 139) of the reign of Emperor Shun of the Eastern Han Dynasty in capital city Luoyang.

Between 17 and 23 years old (AD 94–100), Zhang Heng studied in Changan, the capital city of the Western Han Dynasty (206 BC–AD 8), and Luoyang. At age of 23 (AD 100), he accepted the invitation of the governor of city Nanyang, Bao De (鮑德), as a clerk to manage official documents in his hometown and to assist Bao De in government affairs.

Between 31 and 34 years old (AD 108–111), Zhang Heng stayed in his hometown and studied hard. He specialized in Tai Xuan Jing 《太玄經》 (The Book of the Supreme Mystery) by Yang Xiong (楊雄), which was a philosophical work on cosmic phenomena that discussed astronomy, calendar calculation, and spherical heaven theory.

Zhang Heng was called to the capital city to serve as a palace attendant when he was 34 (AD 111). He became assistant minister at 37 (AD 114). At 38 (AD 115), he became a grand scribe responsible for observing astronomical phenomena, preparing calendars, and managing time devices. At 40 (AD 117), Zhang Heng constructed the armillary sphere. At age 41 he published Ling Xian 《靈憲》 , which was a summary of astronomical theories during his time. The book contained discussions on the evolution of heaven and earth, the cosmos, and the theory of planetary movements. It also contained accurate data on star observations and scientific explanations of the lunar eclipse. At age 42 (AD 119), he wrote the book Suan Wang Lun《算罔論》, a collection of works on the general theory of mathematics. Unfortunately, the book was lost over the passage of time. Zhang Heng also used asymptotic fractions and calculated the ratio of a circle's circumference to its diameter (π) to be the square root of 10, the value of which was between 3.1466 and 3.1622. At age 44 (AD 121), he was made prefect of official carriages, responsible for protecting the imperial palace, transmitting written reports to the emperor, collecting tributes from officials and the public, and receiving envoys to the capital city. Later, he designed and constructed the south- pointing chariot and hodometer. At age 49 (AD 126), Zhang Heng again became a grand scribe and wrote the article Ying Jian《應間》in response to the indifference of the ruling class and the ridicule of the traditionally influential. At age 55 (AD 132), Zhan Heng invented the Di Dong Yi (地動儀), a device for detecting the source of earthquakes. At age 56 (AD 133), he became a palace attendant who served as a consultant and adviser to the emperor. At age 59 (AD 136), he became a governor administering river channels, and a minister at age 61 (AD 138). He died at the age of 62 (AD 139) while serving as a minister.

Zhang Heng was accomplished in writing poems, rhapsodies, literatures, inscriptions, introductions, imperial mandates, eulogies, and calligraphy, as well as figurine sculpture. His poems such as Lyric Poems on Four Sorrows or Si Chou Shi (四愁詩), Tong Sheng Ge (同聲歌), Ge (歌), and Yuan Pian (怨篇), and rhapsodies such as Wen Quan Fu (溫泉賦), Rhapsody of the Two Metropolis or Liang Jung Fu (二京賦), Southern Capital Rhapsody or Nan Du Fu (南都賦), Rhapsody on Contemplating the Mystery or Si Xuan Fu (思玄賦), Rhapsody on Returning to the Fields or Gui Tian Fu (歸田賦), Du Lou Fu (髑髏賦), Zhong Fu (塚賦), Wu Fu (舞賦), Yu Lie Fu (羽獵賦), Ding Qing Fu (定情賦), and Hong Fu (鴻賦), have unique places and values in the history of Chinese literature. Zhang Heng was also one of the twelve famous painters of the Han Dynasty (206 BC–AD 220). Furthermore, Zhang Heng could also draw maps. Not only could he draw the locations of major mountains and rivers all over China, but he was also able to show the geographical features and customs of the areas.

Zhang Heng had many astonishing achievements in wooden machines. More reliable records of his works include the seismoscope, armillary sphere (渾天儀), mechanical calendar (瑞輪冥莢), and wooden flying device (獨木飛雕). He was referred to as the "sage of woodcrafts" together with Ma Jun (馬鈞) of the period of the Three Kingdoms (AD 220–280).

Zhang Heng was an extremely knowledgeable and learned man. Not only was he a great inventor, engineer and scientist, but also a prolific scholar and artist. He can be referred to as the da Vinci of ancient China.

5.4 Historical Records

China has always been plagued by earthquakes. Dynastic histories detailed numerous earthquakes over the centuries. Ancient Chinese emperors were deeply concerned about major earthquakes, since they frequently sparked social unrest in the form of food riots or even rebellions. To maintain control, the government needed to send both food and troops to the suffering region as quickly as possible. Some form of advance warning would therefore have been of great value.

Researches in the relevant literature show that the earliest seismoscope named Hou Feng Di Dong Yi (候風地動儀) was invented by Zhang Heng in the Eastern Han Dynasty (AD 25–220). This instrument was designed to indicate not only the occurrence of an earthquake but also the direction to its source. In the year AD 132, Zhang Heng presented this remarkable device to the Chinese court in the capital Louyang. And in the year AD 138, this device detected an earthquake in Longxi, 400 li, northwest of the capital. The historic records in the Biography of Zhang Heng in the History of the Later Han Dynasty《後漢書·張衡傳》[12] are the most complete ones about Zhang Heng's seismoscope, such as the following description (Figure 5.10): "During the first year of the Yang Chia period (132 AD), Zhang Heng constructed the Hou Feng Di Dong Yi. The instrument was cast with bronze. The outer appearance of it was like a jar with a diameter around eight chi The cover was protruded and it looked like a wine vessel. There were decorations of inscriptions and animals on it. There was a du zhu (a pillar) in the center of the interior and eight transmitting rods near the pillar. There were eight dragons attached to the outside of the vessel, facing in the principal directions of the compass. Below each dragon rested a toad with its mouth open toward the dragon. Each dragon's mouth contained a bronze ball. The intricate mechanism used was hidden inside the device. When the ground moved, the ball located favorably to the direction of ground movement would drop out of the dragon's mouth and fall into the mouth of a bronze toad waiting below. The clang would signify that there had been an earthquake. The direction faced by the dragon that had dropped the ball would be the direction from which the shaking came. And each earthquake only made one ball drop. The device worked accurately. The record showed that one time, the dragon spilled a ball but no earthquake was felt. Scholars in the city thought it was odd. Several days later, news came that an earthquake had indeed occurred in area Longxi. People then realized its ingenuity. From then on, the historian was ordered to record the direction of the quake origins using the device." 『陽嘉元年(西元 132 年)復造候風地動，以精銅鑄成，圓徑八尺，合蓋隆起，形似酒尊，飾以篆文、山龜、鳥獸之形。中有都柱，傍行八道，施關發機；外有八龍，首銜銅丸，下有蟾蜍，張口承之。其牙機巧制，皆隱在尊中，覆蓋周密無際。如有地動，尊則振、龍機發、吐丸，而蟾蜍銜之。振聲激揚，伺者因此覺知。雖一龍發機，比首不動，尋其方面，乃知震之所在。驗之以事，合契若神。自書典所記，未之有也。當一龍機發，而地不覺動，京師學者咸怪其無征。後數日絳至，果地震隴西，于是皆服其妙。自此以後，乃令史官記地動所以方起。』

Figure 5.10 Description of Zhang Heng's seismoscope in the History of the Later Han Dynasty《後漢書□張衡傳》[12]

However, the records that have passed down through history give a detailed account only of the outside of the instrument, Figure 5.11 [13].

Figure 5.11 External appearance of Zhang Heng's seismoscope [13]

5.5 Central Pillar (Du Zhu)

Ancient sources give few practical details regarding the mechanism inside Zhang Heng's seismoscope, except for noting that inside there was a central pillar named du zhu (都 柱) which was capable of lateral displacement along tracks in eight directions, and so arranged that it would operate a closing and opening mechanism.

The development of ancient seismometers used an inertial system. The value of various types of pendulums for earthquake sensing instruments was realized early. Andrea Bina, in AD 1751, was the first to build an instrument for recording the relative displacement of the ground and a pendulum bob in an earthquake. A successful seismograph of low sensitivity was invented by Filippo Cecchi in AD 1875. Nevertheless, British scientists working in Japan in the 1880s developed the seismograph as a practical research instrument. The accuracy of the seismograph was greatly improved in 1898 when Emil Wiechert introduced a very large pendulum with a viscous damping. Therefore, through the research of seismologists, it is believed that the principle of Zhang Heng's seismoscope and modern seismographs are based on the principle of inertia.

In past years, some scholars tried to reconstruct Zhang Heng's seismoscope [13–18] but failed to achieve the high accuracy and sensitivity described in historical records. In 1936, Wang Zhen-duo (王振鐸) presented a reconstruction design of Zhang Heng's seismoscope [13]. In this design

(Figure 5.12), du zhu is a suspended pendulum that functions as a sensor. Once a seismic wave perturbs the pendulum, the pendulum will trigger the nearby lever mechanism to let go the ball in the corresponding dragon's mouth. In 1963, he presented another reconstruction design of Zhang Heng's seismoscope [16]. In this second design (Figure 5.13), du zhu is an inverted pendulum that functions as the sensor.

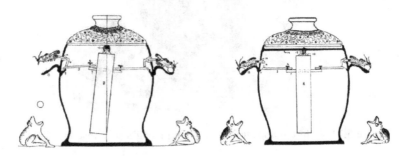

Figure 5.12 First reconstruction design of Zhang Heng's seismoscope by Wang Zhen-duo [13]

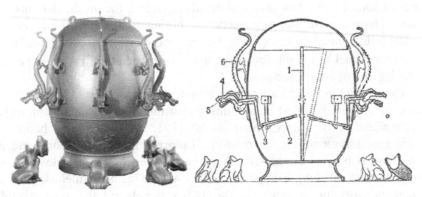

Figure 5.13 Second reconstruction design of Zhang Heng's seismoscope by Wang Zhen-duo [16]

In 1939, Akitsune Imamura built a reconstruction design of Zhang Heng's seismoscope in the Seismological Observatory of Tokyo University [15]. In this design, the diagram follows the slightly modified version by British historian of science, Robert Temple, and worked as follows: "A shock wave from the earth tremor tilted the pendulum so that the spike at the top would swing into one of eight surrounding slots. These contained sliders, the ends of which led into the dragons' mouths. When the spike swung into one of the slots, it would dislodge the slider, which in turn would eject a ball from the dragon's mouth." Trials on an actual model built by Akitsune showed that the device is indeed effective, though in some circumstances the direction of an earthquake's epicenter was found to be at right angles to the dropped ball. In 2006, Feng Rui (馮銳) reported a reconstruction design of Zhang Heng's seismoscope [18]. In this design (Figure 5.14), du zhu is also a suspended pendulum that functions as the sensor.

Figure 5.14   Reconstruction design of Zhang Heng's seismoscope by Feng Rui et al. [18]

It is now generally agreed that the central pillar (du zhu), the key to the device, must have been an inverted pendulum with a weighted bob at the top, ending in a spike that can slide along any one of the eight different channels cut in the surrounding plates. As it enters a channel, it pushes the slider within that channel farther into the dragon's throat, ejecting the ball from the dragon's jaws.

5.6 Design Specifications

The reconstruction of Zhang Heng's seismoscope requires exhaustive literature study to define the problem. Familiarity with science theories and technologies of the subject's time period is also important. Based on literature and seismology study, design specifications of Zhang Heng's seismoscope were concluded through the following process: study of historical archives, investigation of seismology, and analysis of ancient western seismographs, Figure 5.15 [2].

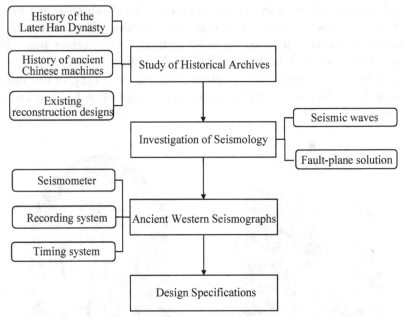

Figure 5.15 Historical developments and design specifications of Zhang Heng's seismoscope

Study of historical archives

There are three components in the study of historical archives relating to seismology: the Biography of Zhang Heng in the History of the Later Han Dynasty, the history of ancient Chinese machines, and existing reconstruction designs.

The Biography of Zhang Heng in the History of the Later Han Dynasty [12] is the most important document on Zhang Heng's seismoscope. The records clearly describe that "there is one pillar (du zhu) in the center of interior and eight transmitting rods near the pillar." This is the first design specification.

The transmitting rod is a channel through which to transport some object. However, no detailed descriptions about the transmitting rod can be found in historical records. Based on the study regarding the development of ancient seismometers, "a basic concept that a switch ball located on the top of the pillar is adapted; and, when an earthquake occurs, the switch ball can move on the transmitting rod." This is the second design specification.

The study of the history of ancient Chinese machines indicated that the developments of levers and linkage mechanisms were very mature and full of various applications, especially in agriculture, military, and textile technology, as presented in Section 3.3. The assembly of a linkage mechanism in ancient China involved the use of revolute joints and prismatic joints. The most elaborate use of linkages in ancient China was in textile machines, in which levers and links were united with treadles to form complicated linkage mechanisms. This understanding of linkage was put to good use in the making of crossbow triggers on military technology. These mechanisms, which involved intricate bent levers and catches, were beautiful, with delicate bronze castings. Available literature shows that the oldest and simplest linkage mechanism that lightened the human labor of dipping, carrying, and emptying buckets was named jie gao, a lever mechanism. This design was familiar before the Qin Dynasty (221–207 BC) in ancient China, and it has been continuously used until the present day. It uses a lever involving rotary motion. A lever arm is supported near its center, weighted with a stone at one end, and loaded by a bucket at the other end, Figure 5.16(a) [19]. The corresponding lever mechanism, the jie gao, consists of a connecting rod and a lever arm, Figure 5.16(b). The joint incident to the lever arm and the ground link is a pin-in-slot joint. The lever arm can slide and rotate around the ground link.

Through the investigation of existing reconstruction designs, the development of the reconstruction of Zhang Heng's seismoscope can be understood. Although the existing reconstruction designs are not so sensitive and accurate, they are useful for understanding the design principle and outer appearance of Zhang Heng's seismoscope.

Investigation of seismology

Seismology includes a lot of subjects. Two topics, seismic waves and fault-plane solution, directly influence the design requirements of Zhang Heng's seismoscope [3–5, 8]. Seismic waves are generated by earthquakes. When an earthquake occurs, seismic waves spread out and travel in specific directions, much like the waves in a pond spreading out from the point where a stone was tossed into the water. There are three kinds of seismic waves, namely P-waves, S-waves, and Surface waves. P-waves are the first-arriving waves from an earthquake. They force the materials in their paths to compress and expand in the direction of wave travel. The direction of P-waves travel is the same as the direction of the earthquake. S-waves are the second-arriving wave from an earthquake. They force particles of materials in their path to move from side-to-side, perpendicular to the wave path. Surface waves are the last-arriving waves that travel along the Earth's surface. An important aspect of the design requirements of Zhang Heng's seismoscope is the first arrival of P-waves. The motion of the ground from the initial P-waves' arrival is known as the first motion. If Zhang Heng's seismoscope can detect the direction of the first motion, the direction of the earthquake can be detected by the instrument.

(a) Jie Gao

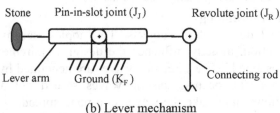

(b) Lever mechanism

Figure 5.16 Structure of jie gao (a lever mechanism) [2]

Although ground shaking can occur from many causes (such as volcanic eruptions), clearly the majority of earthquakes are related to movement along faults. A fault is a shear fracture causing the rock on one or both sides of the fracture surface to slip along it. There are three basic types of faults, namely normal, reverse, and strike-slip fault. Fault-plane solution is a stereographic plot of earthquake-wave first motions that defines the orientation of the fault plane involved, as well as the type of faulting. According to fault-plane solution, the first motion can be either compressing or expanding ground motion. Therefore, "Zhang Heng's seismoscope must detect the direction of the first motion, no matter whether it is compressing or expanding." And, this is the third design specification.

Analysis of ancient western seismographs

Through the analysis of Western seismographs, the developments and design principles of seismographs can be realized [1, 6, 7, 9, 10, 20, 21]. Early seismoscopes were primarily intended to indicate that an earthquake had occurred. A seismoscope designed by J. de la Haute Feuille in AD 1703 contained a central reservoir from which mercury would spill into cups around the periphery if the ground moved. Andrea Bina in AD 1751 proposed a common or clock-type pendulum that was suspended over a tray of sand, so that the pendulum bob could trace a record of ground motion in the sand. Also, scientists in Italy were active in seismic instrument design in the 1800s. The first device was built by Filippo Cecchi in Italy in AD 1875. Cecchi's seismograph used two simple pendulums to measure horizontal motions, one swinging north–south and the other east–west, an orientation still almost always used today. The pendulum motions were magnified three times before being recorded by magnifier. The most significant person in developing a practical seismograph was John Milne. In AD 1876, John Milne, Thomas Gray and J. Alfred Ewing experimented with a variety of pendulum instruments for recording ground motions. In addition, the British in Japan made many observations with their instruments and must be credited with first demonstrating the value to seismology of seismographic devices.

Early seismographs consist of three basic components: a timing system, a recording system, and a seismometer. A clock is used to provide absolute time. A number of different ways have been devised to put time marks on a seismic record. In terms of recording, different methods have been used to record earthquake ground motion, but the most common has been the recording drum. However, the greatest challenge in the development of the seismograph was in the design of a seismometer. A seismometer consists of three basic components: a sensing member, a magnifier, and a long arm. A sensing member responds to ground motion. The motions are magnified before being recorded by a magnifier. The magnifier connects with a long arm as a scribing point which rests on the recording drum. The long arm scribes ground motion in the recording drum. During an earthquake, the ground moves simultaneously in three dimensions: for example, east–west, north–south, and up–down. A single device records only one of these three components of motion. According to the foregoing, it is concluded that "there are eight devices in Zhang Heng's seismoscope to detect eight principal directions. Each device has an interior mechanism as a seismometer inside and a recording system outside." And, this is the fourth design specification.

According to historical records, each recording system of Zhang Heng's seismoscope definitely includes a dragon, a ball, and a toad on the outside. But in the view of the function, the real dragon is not necessary in the recording system. The function of the dragon is to hold the ball. The real dragons are replaced by painting the dragons on the surface of the compass. The balls can be contained in the wall of the vessel. In the developments of seismographs, one or more lever mechanisms were generally used in the early magnifiers of ancient Western seismometers. The most popular lever mechanism in ancient China is jie gao, i.e., a lever mechanism shown in Figure 5.16. Therefore, it is concluded that "each interior mechanism has a pillar as the ground link, a sensing link to respond to ground shake, a lever mechanism (jie gao, including a connecting rod and a lever arm) as a magnifier, and a transmitting rod at least. It is a planar mechanism with one degree of freedom." And this is the fifth design specication. The transmitting rod connects the seismometer and the recording system, much like the long arm in a Western seismograph.

5.7 Reconstruction Design

According to historical records, the outer appearance of Zhang Heng's seismoscope is clear. The reconstruction design presented here focuses on the interior mechanism. The procedure for the reconstruction of design concepts of possible interior mechanisms of Zhang Heng's seismoscope is shown in Figure 5.17. It consists of the following four steps:

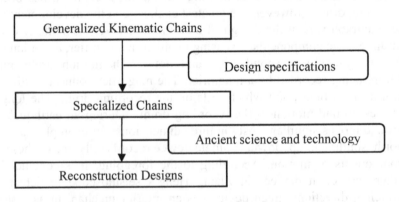

Figure 5.17 Process of reconstruction design

Step 1. Design specifications

Based on the above concluded design specifications, one complete interior mechanism is defined as including a ground link, a sensing link, a connecting rod, a lever arm, and a transmitting rod. The design is at least a planar five-bar mechanism with six joints including one pin-in-slot joint (J_J), one prismatic joint (J_P), and four revolute joints (J_R). This is the simplest structure to carry out the third design specification. In summary, the design requirements of Zhang Heng's seismoscope are:

1. It has one central pillar (du zhu) as the frame in the center of the interior, and it has eight transmitting rods as channels near the pillar.

2. The switch ball which can move on the transmitting rod is held with the eight transmitting rods on the top of the pillar.

3. Zhang Heng's seismoscope must detect the first motion of P-waves; no matter if it is compressing or expanding.

4. There are eight devices in the eight principal directions of Zhang Heng's seismoscope. Each device has the interior mechanism as a seismometer and a recording system.

5. Each interior mechanism has at least a ground link, a sensing link, a connecting rod, a lever arm, and a transmitting rod. It is a planar mechanism with one degree of freedom.

   (a) The pillar in the center of the interior is the ground link. The switch ball is on the top of the ground link.

   (b) The sensing link detects the first motion of P-waves, no matter whether it is compressing or expanding. When the first motion is compressing, the sensing link topples in the direction of its source. On the contrary, if the first motion is expanding, the sensing link topples in the direction away from its source.

   (c) Minimally, the magnifier consists of a connecting rod and a lever arm. Since the joint between the lever arm and the ground link is a pin-in-slot joint, the lever arm can slide and rotate around the pin. The feature of the pin-in-slot joint enables the lever arm to move in the direction of an earthquake. The purpose of such design is to make sure that the movement of the lever arm can follow its corresponding sensing link, no matter where the sensing link topples.

   (d) The function of the transmitting rod is to connect the seismometer and the recording system. When the lever arm moves, it pulls the transmitting rod up. The switch ball drops out of the pillar and moves to the wall of the vessel by the transmitting rod. Through the collision between the balls, the ball in the wall drops out and falls into the mouth of a toad below. The direction of the earthquake is indicated by the dropping ball.

6. It is a planar mechanism with one degree of freedom.

Step 2.   Generalized kinematic chains

The second step is to obtain or identify the atlas of generalized kinematic chains with the required numbers of links and joints subject to defined design specifications (topological characteristics) by applying the algorithm of number synthesis [22] or simply by identifying from Section 4.4.

Step 3.   Specialized chains

The third step is to have the atlas of specialized chains with assigned types of links and joints subject to the concluded design requirements and constraints for each generalized kinematic chain obtained in Step 2. Based on the process of specialization, all possible specialized chains can be identified according to the following substeps:

1. For each generalized kinematic chain, identify the ground link.

2. For each case obtained in substep 1, identify the sensing link.

3. For each case obtained in substep 2, identify the transmitting rod.

4. For each case obtained in substep 3, identify the connecting rod.

5. For each case obtained in substep 4, identify the lever arm.

Specialized chains are identified subject to the following design requirements and constraints. The design constraints are defined based on the concluded characteristics of the interior mechanism.

Ground Link – link (K_F)

1. In each generalized kinematic chain, there must be one ground link (link K_F) as the frame.

2. The ground link must be a link with multiple joints.

Sensing link (link 2)

1. The sensing link (link 2) is adjacent to ground link (link K_F) with a revolute joint (J_R).

2. It must be a binary link.

Transmitting rod (link 5, the channel of the switch ball)

1. The transmitting rod (link 5) is adjacent to ground link (link K_F) with a prismatic joint (J_P).

2. It must be a binary link.

Magnifier (links 3 and 4)

1. The connecting rod (link 3) must be a binary link.
2. It (link 3 or 4) must be adjacent to link K_F.

Step 4. Reconstruction designs

The last step is to obtain the atlas of reconstruction designs from the atlas of specialized chains according to the motion and function requirements of the ancient machinery, and by utilizing the mechanical evolution and variation theory to perform a mechanism equivalent transform. Ancient scientific theories and technologies of the subject's time period are applied to find appropriate and feasible mechanisms that can be considered as the recons-the reconstruction designs.

5.8   Linkage Mechanisms with Five Members

In this case (Example 5.1), there are two generalized kinematic chains with five members and six joints as shown in Figure 4.11 or 5.18(a) and (b). All possible feasible specialized chains are identified through the following steps.

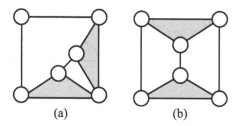

Figure 5.18   Atlas of generalized kinematic chains with five members and six joints

Ground link – link (K_F)

Since there must be a multiple link as the frame, the ground link K_F can be identified as follows:

1. For the generalized kinematic chain shown in Figure 5.18(a), the assignment of the ground link K_F generates one nonisomorphic result, Figure 5.19(a_1).
2. For the generalized chain shown in Figure 5.18(b), the assignment of the ground link K_F generates one nonisomorphic result, Figure 5.19(a_2).

Therefore, two specialized chains with one identified ground link K_F are available as shown in Figure 5.19(a_1) and (a_2).

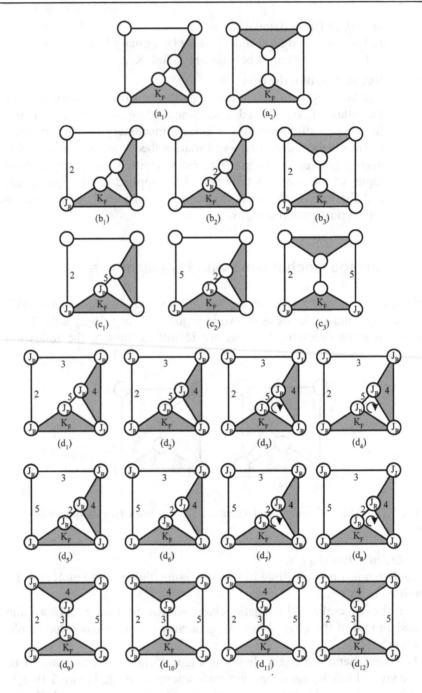

Figure 5.19 Atlas of specialized chains (Example 5.1)

Sensing link (link 2)

Since there must be a binary link as the sensing link 2 that is adjacent to the ground link K_F with a revolute joint J_R, the sensing link can be identified as follows:

1. For the case shown in Figure 5.19(a_1), the assignment of the sensing link 2 generates two results, Figure 5.19(b_1) and (b_2).
2. For the case shown in Figure 5.19(a_2), the assignment of the sensing link 2 generates one nonisomorphic result, Figure 5.19(b_3).

Therefore, three specialized chains with identified ground link K_F and sensing link 2 are available as shown in Figure 5.19(b_1)–(b_3).

Transmitting rod (link 5)

Since there must be a binary link as the transmitting rod 5 that is adjacent to the ground link K_F with a prismatic joint J_P, the transmitting rod can be identified as follows:

1 For the case shown in Figure 5.19(b), the assignment of the transmitting rod 5 generates one result, Figure 5.19(c_1).
2 For the case shown in Figure 5.19(b_2), the assignment of the transmitting rod 5 generates one result, Figure 5.19(c_2).
3 For the case shown in Figure 5.19(b_3), the assignment of the transmitting rod 5 generates one nonisomorphic result, Figure 5.19(c_3).

Therefore, three specialized chains with identified ground link K_F, sensing link 2, and transmitting rod 5 are available as shown in Figure 5.19(c_1)–(c_3).

Magnifier (links 3 and 4)

Since there must be a binary link as the connecting rod 3 and a ternary link as the lever arm 4, the connecting rod and the lever arm can be identified as follows:

1. For the case shown in Figure 5.19(c_1), the assignment of the connecting rod 3, the lever arm 4, the pin-in-plot joint J_J, and the remaining revolute joints J_R generate four results, Figure 5.19(d_1)–(d_4).
2. For the case shown in Figure 5.19(c_2), the assignment of the connecting rod 3, the lever arm 4, the pin-in-plot joint J_J, and the remaining revolute joints J_R generate four results, Figure 5.19(d_5)–(d_8).
3. For the case shown in Figure 5.19(c_3), the assignment of the connecting rod 3, the lever arm 4, the pin-in-plot joint J_J, and the remaining revolute joints J_R generate four results, Figure 5.19(d_9)–(d_{12}).

Therefore, 12 nonisomorphic specialized chains with identified ground link K_F, sensing link 2, transmitting rod 5, connecting rod 3, and lever arm 4 are available as shown in Figure 5.19(d_1)–(d_{12}). Removing those with rigid chains including Figure 5.19(d_3), (d_4), (d_7), and (d_8), eight feasible specialized chains are available as shown in Figure 5.19(d_1), (d_2), (d_5), (d_6), and (d_9)–(d_{12}).

Here, the motion and function requirements of mechanisms are taken into account, and the types of links and joints remained unchanged. Subsequently, the interior mechanism of the corresponding eight feasible specialized chains in Figure 5.19(d_1), (d_2), (d_5), (d_6), and (d_9)–(d_{12}) are represented as shown in Figure 5.20(a)–(h), respectively.

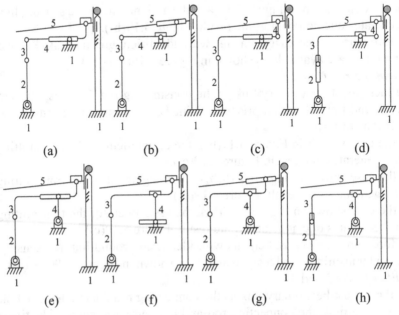

Figure 5.20 Interior mechanisms (Example 5.01)

Figure 5.21 shows the 3D solid model of a reconstruction design of Zhang Heng's seismoscope. There are eight devices in the principal directions of the instrument, Figure 5.21(a). A switch ball is held with the eight transmitting rods (link 5) on the top of the pillar. A complete interior mechanism, based on Figure 5.20(a), is shown in Figure 5.21(b). The sensing link (link 2) detects the first motion of P-waves, no matter whether it is compressing or expanding. When the first motion is compressing, the sensing link 2 topples to the left, Figure 5.21(c). On the contrary, if the first motion is expanding, the sensing link 2 topples to the right, Figure 5.21(d). Figure 5.22 shows the movement process of Zhang Heng's seismoscope.

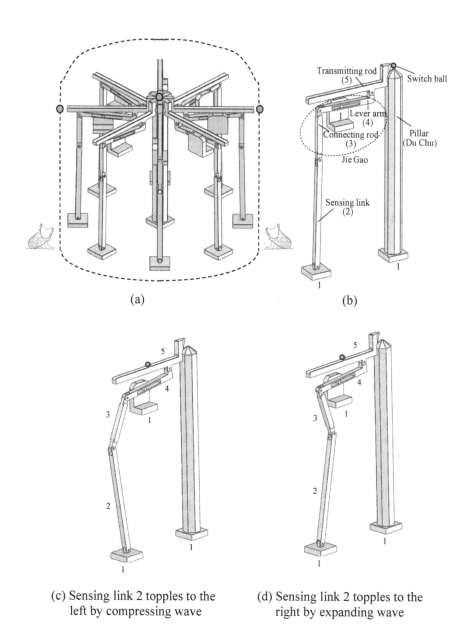

(a)

Transmitting rod
(5)
Switch ball

Lever arm
(4)

Connecting rod
(3)

Jie Gao

Pillar
(Du Chu)

Sensing link
(2)

(b)

(c) Sensing link 2 topples to the
left by compressing wave

(d) Sensing link 2 topples to the
right by expanding wave

Figure 5.21 A reconstruction design of Zhang Heng's seismoscope (Example 7.1)

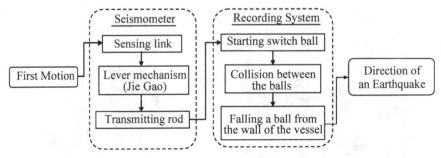

Figure 5.22 Movement process of Zhang Heng's seismoscope

5.9   Linkage Mechanisms with Six Members

In this case (Example 5.2), the interior linkage mechanisms with six members and eight joints can be synthesized by following the same approach as described in Example 5.1. The design requirements and constraints that are different from the (5, 6) interior mechanisms in Example 5.1 are:

1. The magnifier includes links 3–5.
2. The ground link (link K_F) must be a quaternary link.
3. The (6, 8) interior mechanism consists of a ground link (link K_F), a sensing link (link 2), a connecting rod (link 3), two lever arms (link 4 and link 5), a transmitting rod (link 6), a prismatic joint (J_P), two pin-in-slot joints (J_J), and five revolute joints (J_R).
4. Two pin-in-slot joints (J_J) must not be simultaneously incident with the same link.

There are nine generalized kinematic chains with six members and eight joints as shown in Figure 4.14 or Figure 5.23(a)–(i), and all possible specialized chains are identified through the following steps.

Since the interior mechanisms with six members and eight joints must have one sensing link (link 2), one transmitting rod (link 6), and one connecting rod (link 3), a generalized kinematic chain should have only three binary links. Therefore, only those three generalized kinematic chains shown in Figure 5.23(a)–(c) qualify for the process of specialization.
Ground link (link K_F)
Since there must be a quaternary link as the frame, the ground link K_F can be identified as follows:

1. For the generalization chain shown in Figure 5.23(a), the assignment of the ground link K_F generates one result, Figure 5.24(a).
2. For the generalization chain shown in Figure 5.23(b), the assignment of the ground link K_F generates one result, Figure 5.24(b).

3. For the generalization chain shown in Figure 5.23(c), the assignment of the ground link K_F generates one result, Figure 5.24(c).

Therefore, three specialized chains with one identified ground link K_F are available as shown in Figure 7.24(a)–(c).

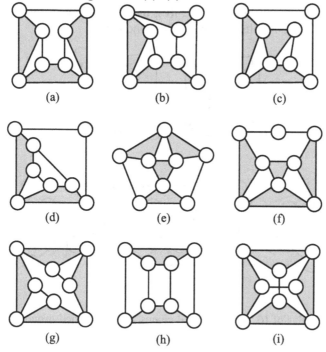

(a) (b) (c)

(d) (e) (f)

(g) (h) (i)

Figure 5.23 Atlas of generalized kinematic chains with six members and eight joints

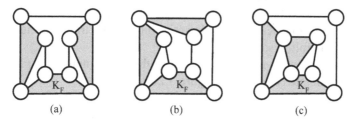

(a) (b) (c)

Figure 5.24 Atlas of (6, 8) specialized chains with identified ground link K_F (Example 5.2)

Sensing link (link 2)

Since there must be a binary link as the sensing link 2 that is adjacent to the ground link K_F with a revolute joint J_R, the sensing link can be identified as follows:

1. For the case shown in Figure 5.24(a), the assignment of the sensing link 2 generates one nonisomorphic result, Figure 5.25(a).
2. For the case shown in Figure 5.24(b), the assignment of the sensing link 2 generates two nonisomorphic results, Figure 5.25(b) and (c).
3. For the case shown in Figure 5.24(c), the assignment of the sensing link 2 generates two results, Figure 5.25(d) and (e).

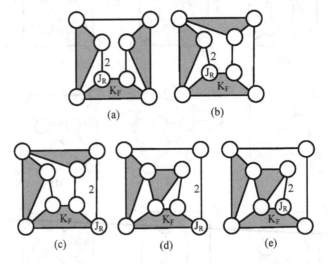

Figure 5.25 Atlas of (6, 8) specialized chains with identified ground link K_F and sensing link 2 (Example 5.2)

Therefore, five specialized chains with identified ground link K_F and sensing link 2 are available as shown in Figure 5.25(a)–(e).

Transmitting rod (link 6)

Since there must be a binary link as the transmitting rod 6 that is adjacent to the ground link K_F with a prismatic joint J_P, the transmitting rod can be identified as follows:

1. For the case shown in Figure 5.25(a), the assignment of the transmitting rod 6 generates one result, Figure 5.26(a).
2. For the case shown in Figure 5.25(b), the assignment of the transmitting rod 6 generates one nonisomorphic result, Figure 5.26(b).
3. For the case shown in Figure 5.25(c), the assignment of the transmitting rod 6 generates two results, Figure 5.26(c) and (d).
4. For the case shown in Figure 5.25(d), the assignment of the transmitting rod 6 generates one result, Figure 5.26(e).
5. For the case shown in Figure 5.25(e), the assignment of the transmitting rod 6 generates one result, Figure 5.26(f).

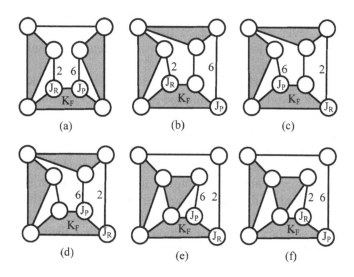

Figure 5.26 Atlas of (6, 8) specialized chains with identified ground link K_F, sensing link 2, and transmitting rod 6 (Example 5.2)

Therefore, six specialized chains with identified ground link K_F, sensing link 2, and transmitting rod 6 are available as shown in Figure 5.26(a)–(f). *Magnifier* (links 3–5)

Since there must be a binary link as the connecting rod 3, the connecting rod and two lever arms 4 and 5 can be identified as follows:

1. For the case shown in Figure 5.26(a), the assignment of the connecting rod 3, two lever arms 4 and 5, two pin-in-plot joints J_J, and the remaining revolute joints J_R generate three feasible results, Figure 5.27(a)–(c).

2. For the case shown in Figure 5.26(b), the assignment of the connecting rod 3, two lever arms 4 and 5, two pin-in-plot joints J_J, and the remaining revolute joints J_R generate five feasible results, Figure 5.27(d)–(h).

3. For the case shown in Figure 5.26(c), the assignment of the connecting rod 3, two lever arms 4 and 5, two pin-in-plot joints J_J, and the remaining revolute joints J_R generate five feasible results, Figure 5.27(i)–(m).

4. For the case shown in Figure 5.26(d), the assignment of the connecting rod 3, two lever arms 4 and 5, two pin-in-plot joints J_J, and the remaining revolute joints J_R generate seven feasible results, Figure 5.27(n)–(t).

5. For the case shown in Figure 5.26(e), the assignment of the connecting rod 3, two lever arms 4 and 5, two pin-in-plot joints J_J, and the remaining revolute joints J_R generate three feasible results, Figure 5.27(u)–(w).

6. For the case shown in Figure 5.26(f), the assignment of the connecting rod 3, two lever arms 4 and 5, two pin-in-plot joints J_J, and the remaining revolute joints J_R generate three feasible results, Figure 5.27(x)–(z).

Therefore, 26 specialized chains with identified ground link K_F, sensing link 2, transmitting rod 6, connecting rod 3, and two lever arms 4 and 5 are available as shown in Figure 5.27(a)–(z).

Through particularizing the 26 feasible specialized chains in Figure 5.27, the corresponding 26 interior mechanisms are obtained as shown in Figure 5.28(a)–(z). Figure 5.29 shows the 3D solid model of an interior mechanism with six members. A complete interior mechanism is shown in Figure 5.29(a) based on the design shown in Figure 5.28(u). When the first motion is compressing, the sensing link (link 2) topples to the left, Figure 5.29(b). On the contrary, if the first motion is expanding, the sensing link 2 (link 2) topples to the right, Figure 5.29(c).

5.10   Rope-and-Pulley Mechanisms with Six Members

In this case (Example 5.3), interior mechanisms with a rope-and-pulley and with six members and eight joints can be synthesized by following the same approach as described in Example 5.2. This design consists of a ground link (link K_F), a sensing link (link 2), a pulley (link 3), a rope (link 4), a lever arm (link 5), a transmitting rod (link 6), a prismatic joint (J_P), a wrapping joint (J_W), a pin-in-slot joint (J_J), and five revolute joints (J_R).

Specialized chains are identified subject to the following design requirements and constraints:

Ground link
1. In each generalized kinematic chain, there must be one ground link as the frame (K_F).
2. The ground link must be a quaternary link.

Sensing link
1. The sensing link is adjacent to the ground link with a revolute joint (J_R).
2. It must be a binary link.

Transmitting rod
1. The transmitting rod, channel of the switch ball, is adjacent to the ground link with a prismatic joint (J_P).
2. It must be a binary link.

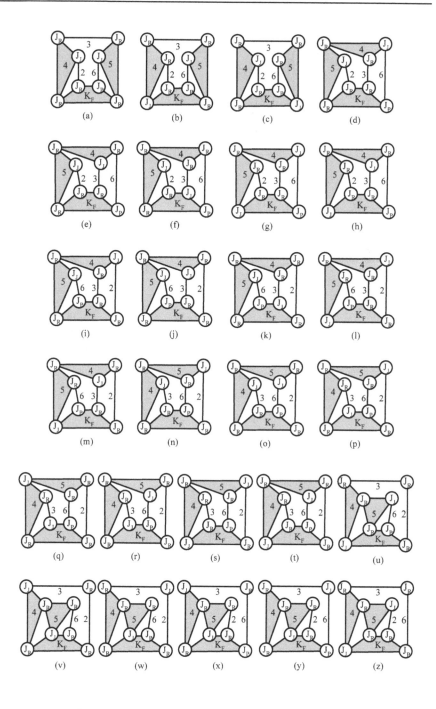

Figure 5.27 Atlas of specialized chains (Example 5.2)

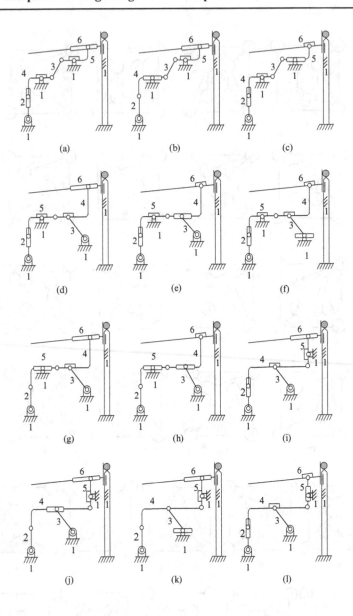

Figure 5.28 Interior mechanisms (Example 5.2)

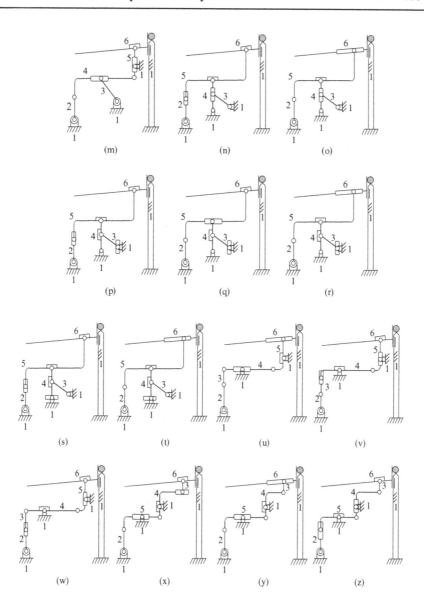

Figure 5.28 (*Continued*)

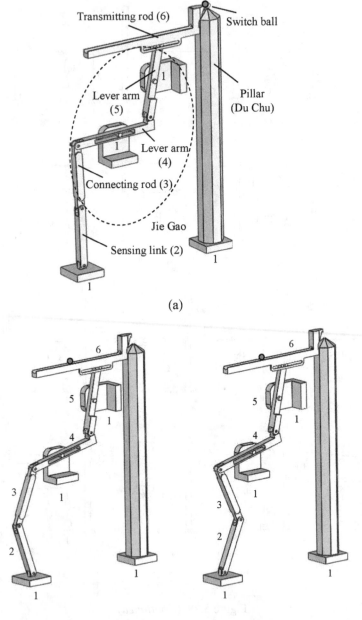

(a)

(b) Sensing link 2 topples to the
left by compressing wave

(c) Sensing link 2 topples to the
right by expanding wave

Figure 5.29 A reconstruction design of Zhang Heng's seismoscope (Example 5.2)

Pulley
1. The pulley is adjacent to the ground link with a revolute joint (J_R).
2. It must be a binary link.
Rope
1. The rope is adjacent to the pulley with a wrapping joint (J_W).
2. It must be a ternary link.
3. The pin-in-slot joint (J_J) must not be incident to the rope.

There are nine generalized kinematic chains with six members and eight joints as shown in Figure 4.14, and again in Figure 5.23(a)–(i).

Since the interior mechanisms with six members and eight joints must have a sensing link, a transmitting rod, and a pulley, a feasible generalized kinematic chain should have only three binary links that are adjacent to the ground link K_F. There must also be a ternary link as the rope. Therefore, only the generalized kinematic chain shown in Figure 5.23(b) qualifies for the process of specialization.

Ground link (link K_F)
Since there must be a quaternary link as the frame, the ground link K_F can be identified as follows:
1. For the generalized kinematic chain shown in Figure 5.23(b), the assignment of the ground link K_F generates one result, Figure 5.30(a).

Therefore, one specialized chain with one identified ground link K_F is available as shown in Figure 5.30(a).

Sensing link (link 2)
Since there must be a binary link as the sensing link 2 that is adjacent to the ground link K_F with a revolute joint J_R, the sensing link can be identified as follows:
1. For the case shown in Figure 5.30(a), the assignment of the sensing link 2 generates two nonisomorphic results, Figure 5.30(b) and (c).

Therefore, two specialized chains with identified ground link K_F and sensing link 2 are available as shown in Figure 5.30(b) and (c).

Transmitting rod (link 6)
Since there must be a binary link as the transmitting rod 6 that is adjacent to the ground link K_F with a prismatic joint J_P, the transmitting rod can be identified as follows:
1. For the case shown in Figure 5.30(b), the assignment of the transmitting rod 6 generates one nonisomorphic result, Figure 5.30(d).
2. For the case shown in Figure 5.30(c), the assignment of the transmitting rod 6 generates two results, Figure 5.30(e) and (f).

Therefore, three specialized chains with identified ground link K_F, sensing link 2, and transmitting rod 6 are available as shown in Figure 5.30(d)–(f).

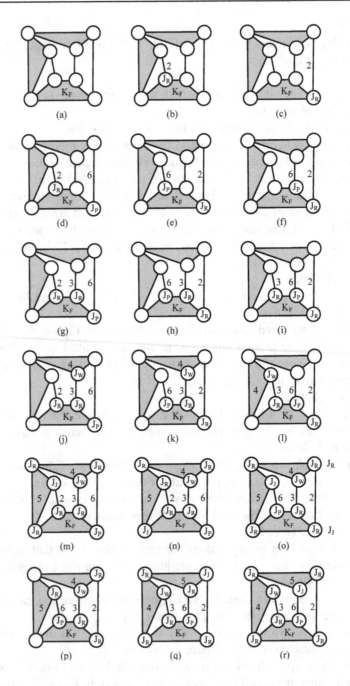

Figure 5.30 Atlas of specialized chains (Example 5.03)

Pulley (link 3)

Since there must be a binary link as the pulley 3 that is adjacent to the ground link K_F with a revolute joint J_R, the pulley can be identified as follows:

1. For the case shown in Figure 5.30(d), the assignment of the pulley 3 generates one result, Figure 5.30(g).
2. For the case shown in Figure 5.26(e), the assignment of the pulley 3 generates one result, Figure 5.30(h).
3. For the case shown in Figure 5.30(f), the assignment of the pulley 3 generates one result, Figure 5.30(i).

Therefore, three specialized chains with identified ground link K_F, sensing link 2, transmitting rod 6, and pulley 3 are available as shown in Figure 5.30(g)–(i).

Rope (link 4)

Since there must be a ternary link as the rope 4 that is adjacent to the pulley 3 with a wrapping joint J_W, the rope can be identified as follows:

1. For the case shown in Figure 5.30(g), the assignment of the rope 4 generates one result, Figure 5.30(j).
2. For the case shown in Figure 5.30(h), the assignment of the rope 4 generates one result, Figure 5.30(k).
3. For the case shown in Figure 5.30(i), the assignment of the rope 4 generates one result, Figure 5.30(l).

Therefore, three specialized chains with identified ground link K_F, sensing link 2, transmitting rod 6, pulley 3 and rope 4 are available as shown in Figure 5.30(j)–(l).

Lever arm (link 5)

Since there must be a lever arm, the lever arm can be identified as follows:

1. For the case shown in Figure 5.30(j), the assignment of the lever arm 5, the pin-in-slot joint J_J, and the remaining revolute joints J_R generates two results, Figure 5.30(m) and (n).
2. For the case shown in Figure 5.30(k), the assignment of the lever arm 5, the pin-in-slot joint J_J, and the remaining revolute joints J_R generates two results, Figure 5.30(o) and (p).
3. For the case shown in Figure 5.30(l), the assignment of the lever arm 5, the pin-in-slot joint J_J, and the remaining revolute joints J_R generates two results, Figure 5.30(q) and (r).

Therefore, six feasible specialized chains with identified ground link K_F, sensing link 2, transmitting rod 6, pulley 3, rope 4, and lever arm 5 are available as shown in Figure 5.30(m)–(r).

Figure 5.31(a)–(f) show the corresponding six interior mechanisms after particularization for the six feasible specialized chains shown in Figure 5.30(m)–(r). Figure 5.32 shows the 3D solid model of an interior mechanism with six members and eight joints. A complete interior mechanism is also shown in Figure 5.32(a) based on the design shown in Figure 5.31(c). When the first motion is compressing, the sensing link (link 2) topples to the left, Figure 5.32(b). On the contrary, if the first motion is expanding, the sensing link 2 (link 2) topples to the right, Figure 5.31(c).

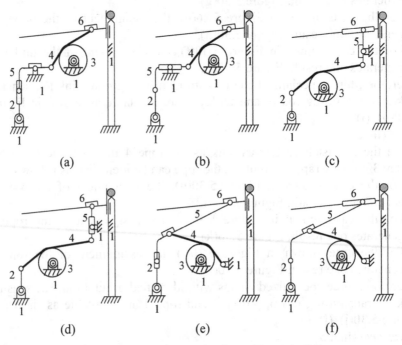

Figure 5.31 Interior mechanisms (Example 5.3)

5.11 Remarks

Earthquakes are natural disasters in the history of mankind. The development of instruments to measure and record the movements of the ground during a distant earthquake in ancient times was truly a great intellectual achievement. The earliest seismoscope was invented by Zhang Heng (張衡) in ancient China in AD 132. The first western seismograph was invented in AD 1703 by the French scientist J. de la Haute Feuile, who knew nothing of his great Chinese predecessor. The first successful modern seismograph was built by Filippo Cecchi in Italy in AD 1875.

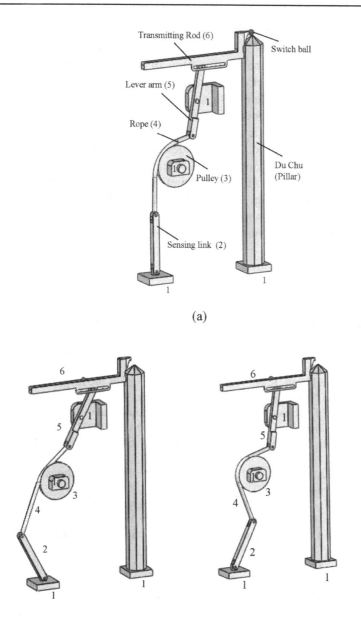

(a)

(b) Sensing link 2 topples to the
left by compressing wave

(c) Sensing link 2 topples to the
right by expanding wave

Figure 5.32 A reconstruction design of Zhang Heng's seismoscope (Example 5.3)

Like many other Chinese inventions, Zhang Heng's seismograph, Hou Feng Di Dong Yi (候風地動儀), was forgotten and lost. Due to insufficient literature on the inner design of the instrument, it has been very difficult to reconstruct the interior device of this seismoscope. With the help of the reconstruction design methodology presented in Chapter 4, all feasible design concepts of Zhang Heng's seismoscopes that are consistent with the scientific theories and technologies of the subject time period can be synthesized. Various designs with different types and numbers of members and joints can be obtained by following the proposed approach. Logically, one of the reconstruction designs is possible to be the inner mechanism of Zhang Heng's seismoscope. This provides a logical and feasible solution and approach to the reconstruction design of Zhang Heng's seismoscope until further new and solid evidence is found.

References

1. Yan, H.S. and Hsiao, K.H., "The development of ancient earthquake instruments," Proceedings of ASME 2006 Design Engineering Technical Conferences – the 30th Mechanisms and Robotics Conferences, paper no. DETC2006-99107, Philadelphia, Pennsylvania, 10–13 September 2006.
2. Yan, H.S. and Hsiao, K.H., "Reconstruction design of the lost seismoscope of ancient China," Mechanism and Machine Theory, Vol. 42, 2007.
3. Hough, S.E., Earthshaking Science: what we know (and don't know) about earthquakes. Princeton University Press, Princeton, NJ, 2002.
4. Levy, M. and Salvadori, M.G., Why the Earth Quakes, Norton, New York, 1995.
5. Yeats, R.S., Allen, C.R. and Sieh, K.E., The geology of earthquakes, Oxford University Press, Oxford, 1997.
6. Howell, B.F., An Introduction to Seismological Research: History and Development, Cambridge University Press, New York, 1990.
7. Brumbaugh, D.S., Earthquakes: Science and Society, Prentice Hall, New Jersey, 1999.
8. Bolt, B.A., Earthquakes and Geological Discovery, Scientific American Library, New York, 1993.
9. Dewey, J. and Byerly, P., "The early history of seismometer (to 1900)," Bulletin of the Seismological Society of America, Vol. 59, No. 1, pp. 183–227, 1969.
10. Milne, J. and Lee, A.W., Earthquakes and other Earth Movement, P. Blakiston's Sons, Philadelphia, 1939.
11. Lu, J.Y., History of Chinese Machines (in Chinese), Ancient Chinese Machinery Cultural Foundation (Tainan, Taiwan), Yue Yin Publishing House, Taipei, 2003.

陸敬嚴，中國機械史，中華古機械文教基金會(台南，台灣)，越吟出版社，台北，2003 年。

12. The History of the Later Han Dynasty (in Chinese) by Fan Ye (Eastern Jin Dynasty), Ding Wen Publishing House, Taipei, 1977.
《後漢書》；范曄[晉朝]撰，鼎文出版社，台北，1977 年。

13. Wang, Z.D., "Conjecture of Zheng Heng's Seismoscope," Yenching University Journal of Chinese Studies (in Chinese), Beijing, Vol. 20, pp. 577–586, 1936.
王振鐸，"漢張衡候風地動儀造法之推測"，燕京學報，第 20 卷，北京，第 577–586 頁，1936 年。

14. Milne, J., Earthquakes and other Earth Movements, Appleton, New York, 1886.

15. Imamura, A., "Tokyo and His Seismoscope," Japanese Journal of Astronomy and Geophysics, Vol. 16, pp. 37–41, 1939.

16. Wang, Z.D., Ke Ji Kao Gu Lun Cong (Papers in Technical Archarology) (in Chinese), Cultural Relics Publishing House, Beijing, 1963.
王振鐸，科技考古論叢，文物出版社，北京，1936 年。

17. Lee, Z.C., Tian Ren Gu Yi (in Chinese), Elephant Publishing House, Zhangzhou, China, 1998.
李志超，天人古義，大象出版社，鄭州，1998 年。

18. Feng, R., Tian, K., Shu, T., Wu, Y., Zhu, X., Li, X., and Sun, X., "Scientific Reconstruction of Zhang Heng's Seismometer," Studies in the History of Natural Sciences, Beijing, Vol. 25, Suppl., pp. 53–76, 2006.
馮銳，田凱，朱濤，伍玉霞，朱曉民，李先登，孫賢陵，"張衡地動儀的科學復原"，自然科學史研究，北京，第 25 卷(增刊)，第 53–76 頁，2006 年 12 月。

19. Tian Gong Kai Wu (in Chinese) by Song Ying-xing (Ming Dynasty), Taiwan Commercial Press, Taipei, 1983.
《天工開物》；宋應星[明朝]撰，天工開物，台灣商務印書館，台北，1983 年。

20. Davison, C., The Founders of Seismology, Cambridge University Press, New York, 1927.

21. Milne, J., Earthquakes, Appleton, New York, 1899.

22. Yan, H.S., Creative Design of Mechanical Devices, Springer, Singapore, 1998.

Chapter 6 Su Song's Escapement Regulator

Su Song (蘇頌) of the Northern Song Dynasty invented a water-powered armillary sphere and celestial globe around year AD 1088. This device was working based on a water-powered mechanical clock with an escapement regulator. Literary records are available for this invention, but unfortunately surviving hardware is lacking. However, several reconstruction designs have existed in the past century.

This chapter systematically generates all feasible designs of Su Song's escapement regulator that meet the scientific and technological standards of the subject's time period [1–4]. Historical records of Su Song's clock tower and the escapement regulator are addressed first. Then, the topological structure of an available design is analyzed, and the design constraints are concluded. The process of reconstruction design of Su Song's escapement regulators is also illustrated. Finally, feasible escapement regulators of Su Song's clock tower and modern mechanical clocks are compared.

6.1 Su Song's Clock Tower

A heavenly body itself can be seen as a time-telling system. Mankind learned to know time, e.g., year, season, month, day, and hour, by observing the regular changes in positions of heavenly bodies such as the sun, the moon and some planets. For more effective observations, the ancient scholars cleverly transformed the orbits of the sun and stars into the dials of the sundial and the gnomon so as to achieve an accurate time telling. The clepsydra was another efficient way of measuring time by watching the flow of a fixed water volume pass through a constant cross section of a siphon. A scale on the water container would lower or rise according to the water level, and calibrations on the scale would indicate the time.

In ancient China, the astronomical clock was integrated from the clepsydra and astronomical instrument as an automatic waterpower timekeeping machine. Its water wheel steelyard-clepsydra mechanism was used to obtain intermittent and equal time keeping. It also had a time-telling system with a cam striking mechanism. The astronomical clock was developed to the most complete and fully functional unit until the water-powered armillary sphere and celestial globe named Shui Yun Yi Xiang Tai (水運儀象台) was built by Su Song and Han Gong-lian (韓公廉) in AD 1088 during the Northern Song Dynasty.

This astronomical clock tower was over thirty chi tall, Figure 6.1. On the top was a massive spherical astronomical instrument for observing the stars, constructed from bronze and driven by water power. Inside the tower was a celestial globe, whose movements were synchronized with those of the sphere above, so that the two could be compared constantly. At the front of the tower was a pagoda-like structure of five floors, each with a door through which wooden puppets appeared at regular intervals throughout the day and night. They beat drums, rang bells, and zhengs (鉦, a type of ancient Chinese ring), played stringed instruments, and displayed tablets showing the time. All these figures were operated by the giant clock machine, powered by a huge water wheel with scoops attached to the end of blades into which water dripped from a water clock, causing the machine to advance by one scoop per hour. Figure 6.2 shows the internal structure of this clock tower.

Figure 6.1 Su Song's clock tower [5]

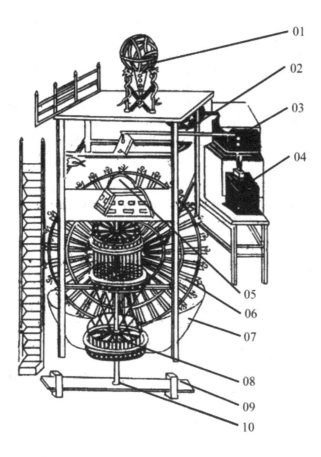

01
02
03
04
05
06
07
08
09
10

1	Armillary sphere (渾儀)
2	Upper balancing lever (天衡)
3	Upper reservoir (天池)
4	Constant-level tank (平水壺)
5	Celestial globe (渾象)
6	Driving wheel (樞輪)
7	Water-withdrawing tank (退水壺)
8	Day and night time-keeping wheels (晝夜機輪)
9	Lower bearing beam (地極)
10	Mortar-shaped end–bearing (樞臼)

Figure 6.2 Internal structure of Su Song's clock tower [5]

Su Song's clock tower reflected ancient Chinese achievements in both astronomy and machinery during the 11th century. In the mechanical field, it was the most outstanding integrated mechanical design, including the water wheel power device, the two-level noria device, the two-level float clepsydra device, the water wheel lever escapement mechanism, and the cam striking time device. The time-telling system of Su Song's clock tower was composed of the day and night time-keeping wheel and the five-storey wooden pagoda. By running the cam striking mechanism, the interaction between concrete image and sound naturally led to three different time laws.

There were differences between the original version and a newer version in the book New Design for an Armillary Sphere and Celestial Globe, named "Xin Yi Xiang Fa Yao" 《新儀象法要》 [5]. The main difference was the transmission system. The transmission from the original version entirely adopted gear trains. The one in the newer version used a mixture of chains and gears. On the day and night time-keeping wheel, the difference between the original and other versions appeared on the mesh position of the transmission gear and the order of their wheels. In the original version, the time-keeping transmission gear was placed on the second layer, which meshed with the middle gear of the transmission shaft. The wheel for puppets reporting the Geng Dian Time Law and the wheel for indicator arrows of the Geng Dian Time Law were placed on the seventh and the eighth layers. In the newer version, the time-keeping transmission gear was put on the fifth layer, which directly meshed with the drum gear of the driving wheel. The places of the wheel for puppets reporting the Geng Dian Time Law and the wheel for indicator arrows of the Geng Dian Time Law were interchanged. Figure 6.3 represents the day and night time-keeping wheel from the other version.

The day and night time-keeping wheel was a six-link design in the time-telling system of Su Song's clock tower. It had two gear joints, two cam joints, and two revolute joints. For the purpose of combining three time laws and using image and sound to tell time automatically, it needed eight layers. They were respectively the celestial transmission gear, the time-keeping transmission gear, the wheel for striking daytime by bell and drum, the wheel for puppets reporting the Duodecimal Time Law, the wheel for puppets reporting the Clepsydra Time Law, the wheel for striking nighttime by zheng, the wheel for puppets reporting the Geng Dian Time Law, and the wheel for indicator arrows of the Geng Dian Time Law. All the above-mentioned eight wheels were installed on the time-keeping shaft together. The top was constrained with an upper bearing beam, and the bottom was supported by an iron mortar-shaped end-bearing.

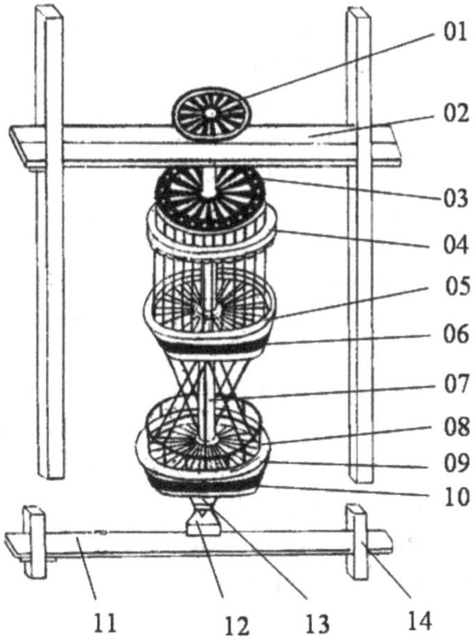

1 Celestial transmission gear (天輪)

2 Upper bearing beam (天束)

3 Wheel for striking daytime by bell and drum （晝時鐘鼓輪）

4 Wheel for puppets reporting the Duodecimal Time Law (時初正司辰輪)

5 Wheel for puppets reporting the Clepsydra Time Law（報刻司辰輪)

6 Time-keeping transmission gear (撥牙機輪)

7 Time-keeping shaft (機輪軸)

8 Wheel for striking nighttime by gong （夜漏金鉦輪）

9 Wheel for puppets reporting the Geng Dian Time Law (夜漏司辰輪)

10 Wheel for indicator-arrows of the Geng Dian Time Law (夜漏箭輪)

11 Lower bearing beam (地極)

12 Mortar-shaped end-bearing （樞臼）

13 Pointed cap of bearing (纂)

14 Base stands (地足)

Figure 6.3 Day and night time-keeping wheel [5]

The time-keeping transmission gear, Figure 6.4(a), was the input terminal of the time-telling system, and meshed with the middle gear of the transmission shaft. It received and transmitted the power and motion of the escapement regulator. The celestial transmission gear, the wheel for striking daytime by bell and drum, and the wheel for striking nighttime by gong, were all output terminals. The celestial transmission gear, Figure 6.4(b), used a gear joint to transmit motive power onto the celestial globe, enabling it to work with astronomical phenomena. The wheel for striking daytime by bell and drum, and the wheel for striking nighttime by zheng, both used a cam joint to operate the striking mechanism on the wooden pagoda to tell time punctually. The joints incident to the time-keeping shaft and the upper bearing beam, as well as the time-keeping shaft and the iron mortar-shaped end-bearing, were both revolute joints. The upper bearing beam was composed of two transverse woods with a semicircle notch as the frame for holding the time-keeping shaft. The mortar-shaped end-bearing was to support the pointed cap, named zuan (纂) of the time-keeping shaft. Both materials were iron. This joint was similar to a self-aligning taper journal bearing.

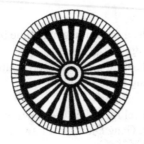 

(a) Time-keeping transmission gear   (b) Celestial transmission gear

Figure 6.4 Transmission gear on the day and night time-keeping wheel [5]

The five-storey wooden pagoda, Figure 6.5, was used to cover the day and night time-keeping wheel. At every storey doors were built through which puppets showed up or vanished, making it a time display platform with interacting image and music.

The first storey of the wooden pagoda consisted of three doors, the left door, middle door, and right door. Three puppets wearing different colored clothes, stood at each of the three doors respectively. Each puppet had an active arm with linkage mechanisms, which was respectively operated by three teams of pegs on the wheel for striking the daytime bell and drum,

Figure 6.6(a). This was a cam striking time-telling device. The time sequence and disposition among the three teams of pegs on the wheel for striking the daytime bell and drum corresponded with one-shi (時) intervals, one-ke (刻) intervals, and intervals of 600 pegs (teeth) on the time-keeping transmission gear, respectively. In this way, at the beginning of each time interval shi, a puppet in scarlet clothes shook a small bell at the left door. At each time interval ke, a puppet wearing green clothes beat the drum at the middle door. At the middle of each time interval shi, a puppet wearing purple struck a large bell at the right door.

Figure 6.5 Five-storey wooden pagoda [5]

The second storey of the wooden pagoda had only one door opened at the middle. The wheel for puppets reporting the Duodecimal Time Law, Figure 6.6(b), was inside the second storey. There were 24 puppets standing by the outer rim of the wheel, each holding a tablet. The beginnings and middles of twelve shi was in written on every tablet. They corresponded to the 600 teeth of the time-keeping transmission gear. Thus at the beginning of each time interval shi, the puppet in scarlet appeared holding the tablet at the door. At the middle of each time interval shi, the puppet in purple appeared holding the tablet at the door.

(a) Wheel for striking daytime by bell and drum

(b) Wheel for puppets reporting the Duodecimal Time Law

(c) Wheel for puppets reporting the Clepsydra Time Law (only 36 puppets shown)

Figure 6.6 Telling daytime devices on the day and night time-keeping wheel [5]

The third storey of the wooden pagoda also had a door opened at the middle. Inside the third storey was the wheel for puppets reporting the Clepsydra Time Law, Figure 6.6(b). There were 96 puppets standing by the outer rim of this wheel. Each of them held a tablet, in turn written 100 ke of twelve shi. They also matched with the 600 teeth of the time-keeping transmission gear. At each time interval ke, a puppet in green appeared holding a tablet at the middle door.

The fourth and the fifth stories both had a door opened right at the middle. Being the night time-telling device, they specialized in reporting the Geng Dian Time Law, together with the wheel for striking nighttime by zheng, the wheel for puppets reporting the Geng Dian Time Law, and the wheel for indicator arrows of the Geng Dian Time Law.

On the fourth storey, a striking zheng puppet stood affixed at the middle door. Its action was similar to the one in the first storey. Poked by the teeth of the wheel for striking nighttime by zheng, Figure 6.7(a), at sunset, at dusk, at each geng (更) and dian (旦), at predawn, at dawn and at sunrise, it responded to the appearance of a puppet on the fifth storey. It was also a cam striking time-telling device.

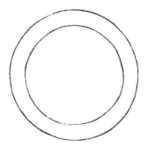 

(a) Wheel for striking nighttime   (b) Wheel for puppets reporting the Geng
     by gong   Dian Time Law (only 12–4 puppets
                                          shown)

Figure 6.7 Telling nighttime devices on the day and night time-keeping wheel [5]

Inside the fifth storey was the wheel for puppets reporting the Geng Dian Time Law, Figure 6.7(b). There were 38 puppets wearing three different colors of clothes. Based on the data of each indicator arrow on the wheel for indicator arrows of the Geng Dian Time Law, the puppets come out to report the Geng Dian Time Law through their relative positions on wheel. Therefore at sunset, a puppet in scarlet appeared to report, and then after 2½ ke another one in green came out to report dusk. During the five geng, there were 25 reports. At each geng a puppet in scarlet appeared to report; at each dan a puppet in green appeared to report. At each ke of predawn a puppet in green appeared to report. At dawn a puppet in green reported, and at sunrise a puppet in scarlet reported. All these puppets appeared in the central doorway.

In the process of restoring Su Song's clock tower in AD 1958, Wang Zhen-duo (王振鐸) took the arrows on the wheel for indicator arrows of the Geng Dian Time Law as the pegs (teeth) on the wheel for striking nighttime by geng [6]. And there were three groups of holes on the outer rim of the wheel for striking nighttime by geng. The division rule of the three group holes was based upon the summer solstice, winter solstice, and spring equinox (or autumn equinox). The holes were also grouped to receive insertions of the indicator arrows corresponding to the four seasons. However, this work caused the same problems. First the mechanism of the wheel for striking nighttime by geng became complicated, and, the wheel for indicator arrows of the Geng Dian Time Law seemed to be an unnecessary device. Second, it seemed too general that the yearly Geng Dian Time Law was only divided into four periods. It was no longer able to reflect the reality of astronomical phenomena, thus becoming extremely discordant with scientific spirits. Nevertheless, some later scholars still adhered to Wang Zhen-duo's restoration model [7–10], even though it was unable to justify itself whenever compared with the original text.

According to the application of Clepsydra Time Law, the wheel for indicator arrows of the Geng Dian Time Law was a database. There were 61 indicator arrows placed on this wheel. The nighttime Geng Dian Time Law fluctuated in its time recording. As the length of the night varied with seasonal changes and solar terms, the 61 arrows were set in a year (365.25 days), and each arrow was used approximately for 6 days. To achieve a better identification, the length of each arrow was designed in proportion to that of the corresponding night. On each arrow was written the time and record of its use, such as the sunrise and the sunset of that period, how many ke and fen (分) were included in each geng, and how many ke and fen were there within each dan. In this case, it was very convenient that the position of the pegs (teeth) on the wheel for striking nighttime by zheng, and the position of the puppets on the wheel for puppets reporting the Clepsydra Time Law, adjusted to seasonal changes according to the data of the corresponding arrow on the wheel for indicator arrows of the Geng Dian Time Law. Moreover, this application not only conformed to the actual record on the original text, but also conformed to the scientific spirit of Su Song's clock tower. In the old record of the clepsydra law, 41 arrows and 48 arrows were successively used. Here the use of 61 arrows was a vital evidence of Su Song's high standard on time precision.

The motion of Su Song's clock tower was controlled by the escapement regulator, which was made up of the time steelyard-clepsydra device and the water wheel lever escapement [2]. The time steelyard-clepsydra device operated through the use of the repetitive accumulation of energy and the periodic release of energy regulated water flow in the two-level float clepsydra. The periodic swing of the clepsydra was thus under control enabling the water wheel lever escapement to maintain a precise and accurate intermittent motion. In this way, the day and night time-keeping wheel could uniformly rotate one turn per day.

The day and night time-keeping wheel rotating one turn per day was a design objective and constraint when Su Song was building this clock tower. So the design of the tooth number of the time-keeping transmission gear became the focus. This entailed some problems, such as the time precision of the whole machine, the sequence of the cam striking device, and the reporting order of the puppets of the time-keeping wheel. With regard to the precision of time keeping, it was the tooth number of the time-keeping transmission gear that determined the precision of the time-telling system. This wheel had 600 teeth. To make one turn in a day, this machine was running up to 100 ke for a full day and night which totaled 6,000 fen. So when the wheel turned one tooth, it went 10 fen. When it turned six teeth, it was 1 ke. The reading precision almost equaled 10 fen (144 s). Hence the time-keeping precision of the escapement regulator of Su Song's clock tower must be higher than 144 s. Such precision mainly depended on the time steelyard-clepsydra device. Derived from the record in the Song Dynasty (AD 960–1279) and the experiment of modern restoration, the reading precision of the clepsydra was proved to be less than 14.4 s; otherwise the time-keeping precision was possible to reach 10^{-4} [2]. This proved the fact that the design and completion of the time-keeping transmission gear possessing 600 teeth was a reasonable device.

6.2 Su Song the Man

A native of Tongan in Quanzhou (currently Tongan in province Fujian), Su Song, also known as Zi Rong (子容), was born during the 4th year of the reign of Emperor Zhen Zong (AD 1020) during the Northern Song Dynasty (AD 960–1126). He died in the 1st year of the reign of Emperor Hui Zong (AD 1101) at age 82 [11].

Su Song was born to a family of imperial officials. His father, Su Shen (蘇紳), attained their rank of Jinshi (進士) metropolitan graduate in the 3rd year of the reign of Emperor Ren Zong (AD 1019), and became a literary attendant during the reign of Emperor Ren Zong. At an early age, Su Song received a fine education and strict training from his father, and grew up to become an assiduous reader. At age 23, Su Song began his career as a public official after he passed the Jinshi examination during the 2nd year of the reign of Emperor Ren Zong (AD 1042). Su Song served as a local official and later served with the imperial government in the capital city. Over the years, he also served in other places on several occasions. During the 1st year of the reign of Emperor Zhe Zong (AD 1086), Su Song was recalled to the capital city to become a minister of justice, a minister of personnel and academician on the waiting list, an academician with the imperial academy reviewing imperial edicts, and a royal secretary. Su Song was a chancellor for 9 months from 7th to 8th year of the reign of Emperor Zhe Zong (AD 1092), after which he resigned. After Su Song left his post as a chancellor, he again served as a local official. During the 4th year of the reign of Emperor Zhe Zong (AD 1097), Su Song, at age 78, finally resigned from public office and lived in Jingkou (now Zhenjiang in province Jiang Su) until his death. Su Song was posthumously promoted to the post of imperial examiner and Wei Guo Gong (魏國公), and bestowed the title of Zheng Jian (正簡). "To maintain fairness in discussions, remain upright in the handling of affairs, avoid having favors with the powerful and influential, and not to organize private hordes" were Su Song's guiding principles [7].

Su Song was a learned and diligent scholar whose academic achievements surpassed his political accomplishments. His major contributions include the compilation of the book Illustrated Canon of Meteria Medica, also named Ben Cao Tu Jing《本草圖經》, the direction in the design and creation of a water-powered armillary sphere and celestial globe namely Shui Yun Yi Xiang Tai (水運儀象台), and authorship of the book New Design for an Armillary Sphere and Celestial Golbe, also named Xin Yi Xiang Fa Yao《新儀象法要》.

Su Song compiled the Illustrated Canon of Materia Medica when he was an official editor of Chinese classics and medical texts. The book was completed in the 6th year of the reign of Emperor Ren Zong (AD 1061). The Illustrated Canon is a compilation of illustrated information on herbal medicines from all over China. Since the explanations in the original version were inconsistent and vulgar, Su Song compiled, sorted, classified, and catalogued the information. In addition, he refined the writing and scrutinized the facts. The Illustrated Canon not only showed the survey of available medicines in China at that time, but also provided information on those from foreign lands. It also contained many historical medicinal literature and medical records from before the Song Dynasty (AD 960–1279). Due to Su Song's effort, the book is China's greatest illustrated compilation of herbal medicines. For this reason, many later literary works often used the Illustrated Canon of Materia Medica as a reference.

The water-powered armillary sphere and celestial globe (Shui Yun Yi Xiang Tai) exemplifies achievement in astronomy and mechanical engineering during the 11th century of ancient China. The armillary sphere was built by Han Gong-lian and other technicians of the astronomical observatory office under the direction of Su Song from AD 1086 to 1092. The armillary sphere tower is a mechanism that integrated the functions of the armillary sphere, celestial globe, and mechanical clock. However, the large-scale water-driven astronomical clock tower was destroyed when the Mongolians invaded the capital city. Fortunately, Su Song had written down the origin and creation process, and had drawn pictures of the entire mechanism as well as its components in his book New Design for an Armillary Sphere and Celestial Globe (Xin Yi Xiang Fa Yao). His work introduced the water-driven armillary sphere to the world. The book contains 63 pictures where 14 are astronomical diagrams and 49 are mechanical drawings. Each illustration contains explanations of the name, dimension, construction, and movement of every component. The book is an important technological reference for later generations, especially in astronomical and mechanical researches.

6.3 Su Song's Escapement Regulator

Literature researches show that the earliest escapement regulators were invented in ancient China. Yet, because literature on the subject is limited, present day scholars have different opinions about when the first escapement regulator was invented and who had invented it [7, 12–14]. Some scholars believe that the original creators were Buddhist monk Yi Xing (一行) and Liang Ling-zan (梁令瓚) in the Tang Dynasty (AD 618–906), while others believe that Zhang Si-xun (張思訓) or Su Song and Han Gong-lian in the Northern Song Dynasty (AD 960–1126) were the first inventors. Furthermore, the article Annals on Metrology and Calendar of the History of Song Dynasty《宋史·律曆志》about the device ji heng (璣衡, an astronomical clock) made by Wang Fu (王黼) in AD 1124 reads: "… the escapement regulator used in jin heng originated from that made by monk Yi Xing during the Tang Dynasty." [15] This passage suggests that although Su Song's water wheel steelyard-clepsydra device was an example of a successful escapement regulator, it was not the first such device invented in ancient China. As for the other inventors, because of incomplete historical records or lack of diagrams, the actual structures of their inventions remain largely unknown. Although in AD 1958, Joseph Needham reaffirmed that "There are enough materials to prove that the first water wheel linkwork escapement mechanism was invented by Yi Xing and Liang Ling-zan in 720 AD" [16], he was still unable to provide a mechanism diagram of the device. These are common difficulties scholars encounter, when exploring ancient machinery, especially in reconstruction research on lost machinery that was documented but lack physical evidence.

Among these possible inventors, only Su Song wrote the book New Design for an Armillary Sphere and Celestial Globe during the period of AD 1088 to 1096 [5], documenting in detail the structure, components, and diagrams of the motion and structure of the water-powered clock tower. The book clearly describes how the time steelyard-clepsydra device and the water wheel lever escapement worked in unison to perform the isochoric and intermittent timekeeping function. The book enabled the escapement regulator using the water wheel and steelyard-clepsydra mechanism to be handed down to future generations.

Su Song's clock tower reflected Ancient Chinese achievements in both astronomy and mechanics during the 11th century. This outstanding mechanical design included the water wheel power device, the two-level noria device, the two-level float device, the water wheel lever escapement mechanism, the programmable cam mechanism, and the time-telling device. The mechanisms and mechanical components used included gears, chains, linkages, ratchets, cams, hinges, and sliding bearings. Among the mechanisms, the escapement regulator made up of the time steelyard-clepsydra device and the water wheel lever escapement (Figure 6.8) already had the functions and capabilities of the escapement regulators of modern mechanical clocks.

The time steelyard-clepsydra device was composed of a two-level float clepsydra device, a water-receiving scoop, a lower balancing lever, a lower weight, and a checking fork. The two-level float device included an upper reservoir and a constant-level tank. The book New Design for an Armillary Sphere and Celestial Globe read [5]: "The constant-level tank had a water-level marker. Water was lifted to a reservoir and poured into the upper reservoir. A constant-level tank was used to regulate water flow to maintain constant the speed and amount of water flowing from the upper reservoir. Water then flowed into the water-receiving scoops on the driving wheel. Since the water flow was maintained constant throughout the day, accurate time measurement was ensured. ... A lower balancing lever and a lower weight were located above the stopping tongue of the upper balancing lever. A free-spinning axle was located at the center of the lower balancing lever, which was held in place by two plates installed at the crossbar located at the north-south direction of the stand holding the constant-level tank. The tip of the lower balancing lever was a checking fork, which alternately checked and released the water-receiving scoops on the driving wheel. The lower weight was located on the opposite end of the lower balancing lever, which would rise or lower itself in accordance with the amount of water inside the water-receiving scoop." 『平水壺上有準水箭，自河車發水入天河，以注天池壺。天池壺受水有多少緊慢不均，故以平水壺節之，即注樞輪受水壺，晝夜停勻時刻自正。... 樞衡、樞權各一，在天衡關舌上，正中爲關軸於平水壺南北橫桄上，爲兩頰以貫其軸，常使運動。首爲格叉，西距樞輪受水壺，權隨於衡東，隨水壺虛實低昂。』

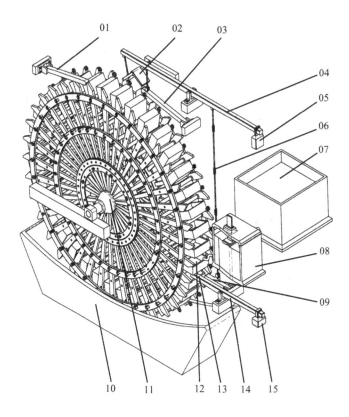

01 Left upper lock (左天鎖)   09 Upper stopping tongue (關舌)
02 Upper stopping device (天關)   10 Water-withdrawing tank (退水壺)
03 Right upper lock (右天鎖)   11 Driving wheel (樞輪)
04 Upper balancing lever (天衡)   12 Water-receiving scoop (受水壺)
05 Upper weight (天權)   13 Checking fork (格叉)
06 Connecting rod (天條)   14 Lower balancing lever (樞衡)
07 Upper reservoir (天池)   15 Lower weight (樞權)
08 Constant-level tank: a cutaway view (平水壺)

Figure 6.8 Escapement regulator of Su Song's clock tower [3]

The water wheel lever escapement was made up of the driving wheel, the left and right upper lock, the upper stopping device, the upper balancing lever, the upper weight, the connecting rod, and the upper stopping tongue. The driving wheel (Figure 6.9) transformed the potential energy from the water level to drive the entire machine, and acted as the escape wheel as well. And, Figure 6.10 shows the upper balancing lever mechanism in Su Song's clock tower.

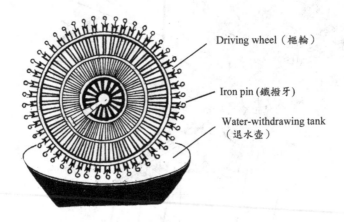

Driving wheel (樞輪)

Iron pin (鐵撥牙)

Water-withdrawing tank
(退水壺)

Figure 6.9 Driving wheel of Su Song's clock tower [5]

The book New Design for an Armillary Sphere and Celestial Globe indicated [5]: "An upper balancing lever was mounted above the driving wheel, with a metal axle installed at the center of the lever. "Camel backs" were mounted on the cross-bar, on which were two metal plates. The metal axle was installed in between these metal plates, thus enabling the upper balancing lever to rotate freely. An upper weight was hanged at one end of the upper balancing lever, while an upper stopping device was connected to the other end. A chain-like connecting rod was connected to the right of the upper balancing lever, between the axle and upper weight. The length of the connecting rod was determined by the position (height) of the driving wheel. There was an upper stopping tongue with a metal axle connected to its end. The axle was mounted on a cross-bar in the north-south direction of the stand holding the constant-level tank, allowing it to rotate freely. The front end of the stopping tongue was connected to the end of the connecting rod; when the stopping tongue turned downward, the upper stopping device would be pulled upward. There were also left and right upper locks, with axles connected at the end of the locks. Axles were installed at the cross-bars on the two left and right posts. The left and right upper locks were installed in opposite directions to check and release the driving wheel." 『天衡一，在樞輪之上中爲鐵關軸於東天柱間橫桄上，爲馳峰。植兩鐵頰以貫其軸，常使轉動。天權一，掛於天衡尾；天關一，掛於腦。天條一（即鐵鶴膝也），綴於權裏右垂（長短隨樞輪高下）。天衡關舌一，末爲鐵關軸，寄安於平水壺架南北桄上，常使轉動，首綴於天條，舌動則關起。左右天鎖各一，末皆爲關軸，寄安左右天柱橫桄上，東西相對以拒樞輪之輻。』

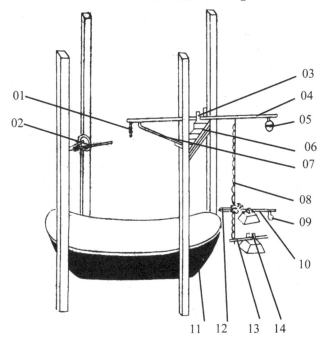

01	Upper stopping device (天關)	08	Connecting rod (天條)
02	Right upper lock (右天鎖)	09	Lower weight (樞權)
03	Camel back (駝峰)	10	Lower balancing lever (樞衡)
04	Upper balancing lever (天衡)	11	Water-withdrawing tank (退水壺)
05	Upper weight (天權)	12	Checking fork (格叉)
06	Cross-bar (橫桄)	13	Upper stopping tongue (關舌)
07	Left upper lock (左天鎖)	14	Shutting axle (關軸)

Figure 6.10 Upper balancing lever mechanism in Su Song's clock tower [5]

The time steelyard-clepsydra device used the repetitive accumulation of energy and the periodic release of energy to regulate water flow in the two-level float clepsydra. The periodic swing of the clepsydra was thus controlled to enable the water wheel lever escapement to maintain a precise and accurate intermittent motion. The result was precision time keeping. The book New Design for an Armillary Sphere and Celestial Globe contained a detailed description of its operation [5]: "Operations of Su Song's clock-tower start from lower water-raising tank. ... Water from the upper reservoir flowed into the constant-level tank through the siphon, where it was fed to the water-receiving scoops on the driving wheel located to the west of the tanks. The bottom of the water-receiving scoops ran counter to the metal checking fork on the lower-balancing lever, and the checking fork was used to check and release the water-receiving scoops. When empty, the water-receiving scoop was locked in place by the checking fork, allowing the scoop to receive water coming from the siphon. When full, the scoop pressed down on the checking fork, and the iron pin at the outer edge of the scoop would disengage the upper stopping tongue located beneath the checking fork, which would pull on the connecting rod. When pulled, the connecting rod raised the front of the upper balancing lever, jerking open the left upper lock and the upper stopping device. This would release the drive wheel, allowing it to turn one spoke due to the imbalance between the empty scoops on the one side and the full scoops on the others. At this point, the upper stopping device would engage with the left upper lock, allowing water from the previous water-receiving scoop to be emptied into the water-withdrawing tank. At the same time, the upper stopping device and the upper lock worked together to lock in place the next water-receiving scoop as it descended. The right upper lock prevented the wheel from recoiling and turning backwards. When the next scoop was released, the water would again be emptied into the water-withdrawing tank. Water in the water-withdrawing tank then flowed into the lower water-raising tank through an opening at the bottom of the tank. The action would then begin another cycle, enabling the astronomical clock to continue its non-stop operation." 『水運之制始於下壺，...天池水南出渴烏，注入平水壺；由渴烏西注，入樞輪受水壺。受水壺之東與鐵樞衡格叉相對，格叉以距受水壺。壺虛，即爲格叉所格，所以能受水。水實，即格叉不能勝壺，故格叉落，格叉落即壺側鐵撥擊開天衡關舌，掣動天條；天條動，則天衡起，發動天衡關；左天鎖開，即放樞輪一輻過；一輻過，即樞輪動。 ... 已上樞輪一輻過，則左天鎖及天關開；左天鎖及天關開，則一受水壺落入退水壺；一壺落，則關、鎖再拒次壺，激輪右回，故以右天鎖拒之，使不能西也。每受水一壺過，水落入退水壺，由下竅北流入昇水下壺。再動河車運水入上水壺，周而復始。 』

6.4   Reconstruction Design

In what follows, the procedure of the reconstruction design methodology shown in Figure 4.1 is extended to generate all possible design concepts of Su Song's escapement regulators.

6.4.1 Design specifications

Escapement regulators are important features of mechanical clocks. They consist of two major parts: an oscillator and an escapement. The oscillator is a device that generates isochronous and periodic motion; the escapement, on the other hand, is a motion-controlling mechanism. The escapement regulator relies on the periodic vibration of the oscillator to maintain accurate and uniform intermittent motion in the escapement; it thus functions as a speed regulator.

The development of ancient Chinese escapement regulators lies in the knowledge of clepsydra and lever technologies. In ancient China, applications of clepsydra and lever mechanisms were ubiquitous, with steady improvements in the structures, forms, and accuracy documented in historical records. The clepsydra, utilizing the steady flow of water from a reservoir and an arrow to indicate time, was the predominant timer used in ancient China. As for their structures, the floating clepsydra and the steelyard-clepsydra were the two major types. There were also improved mercury-driven and sand-driven clepsydras that were used to avoid the defect of water-driven clepsydras. The most popular lever mechanisms in ancient China were the jie gao (桔槔, a labor-saving lever with unequal arms) and heng qi (衡器, a weighing apparatus) as previously presented in Section 3.3. An escapement can be made by integrating the jie gao as a force amplifier and the heng qi as a weight comparator to control the motion of the water wheel.

Thus, the design specifications of a water wheel steelyard-clepsydra device are as follow:
1. It is an escapement regulator.
2. It has a water wheel.
3. It has an independent input that has an isochoric and intermittent motion.
4. It has an escapement that can control the water wheel motion.

According to the result of a literature research, the picture of the water wheel steelyard-clepsydra device used in Su Song's clock tower is clear as shown in Figure 6.8. It was composed of a time steelyard-clepsydra device and a water wheel lever escapement, which was a unique feature of escapement regulators in ancient China.

The time steelyard-clepsydra device generated uniform and periodic motion by integrating a balancing mechanism with a two-level float-clepsydra. It consisted of the following components: an upper reservoir, a constant-level tank, a water-receiving scoop, a lower balancing lever, a lower weight, and a checking fork.

The water wheel lever escapement was composed of a ratchet mechanism and an upper balancing lever. It received the regular momentum produced by the oscillation system, enabling it to generate periodic vibration to check and release the intermittent motion of the driving wheel. Its mechanical members included a driving wheel, left and right upper locks, an upper stopping device, an upper balancing lever, an upper weight, a connecting rod, and an upper stopping tongue.

By repeatedly accumulating and periodically releasing energy, the time steelyard-clepsydra device can be used to regulate the two-level float-clepsydra to a constant speed and to control the oscillating frequency of the steelyard-clepsydra. This is how the water wheel lever escapement maintains the accurate and isochoric intermittent motion necessary to precisely track time.

Figure 6.11 shows the corresponding mechanism sketch of Su Song's water wheel steelyard-clepsydra shown in Figure 6.8. The driving wheel (K_2, a water wheel) is the power wheel of the whole clock tower. It uses a wheel axle to transmit power to the armillary sphere, the celestial globe, and the time-telling device through a gear transmission mechanism. The wheel axle is adjacent to the frame (K_G) with a revolute joint (J_R). The time steelyard-clepsydra device is an oscillator, using the water-receiving scoop (K_6) to output isochoric periodic movement. The water-receiving scoop is adjacent to the driving wheel with a revolute joint. The upper balancing mechanism is a kind of lever mechanism. Its upper stopping tongue (K_5), which receives the isochoric and intermittent momentum from the water-receiving scoop, is adjacent to the frame with a revolute joint. The joint incident to the upper stopping tongue and the scoop is a cam pair (J_A).

The connecting rod (K_4) is similar to a chain. It is a flexible member which transmits long-range motion and power between the upper balancing lever (K_3) and the upper stopping tongue. Its topological structure is equivalent to a rigid link, with revolute joints incident to the upper balancing lever and the upper stopping tongue respectively. The upper balancing lever is adjacent to the frame with a revolute joint and connects with the driving wheel by an upper stopping device. The purpose of such design is to control the movement of the upper balancing lever in an up-and-down order, and thereby acts accurately on the check and release motions of the driving wheel. However, no detailed descriptions about the upper stopping device can be found in historical records. Nevertheless, in the conceptual design stage, it is not necessary to focus on its complete specifications. It is considered as a planar pair with two degrees of freedom, an upper stopping joint denoted as J_T. And, the corresponding topology matrix M_T is:

$$
M_T = \begin{bmatrix}
K_G & J_R & J_R & 0 & J_R & 0 \\
a_0 & K_2 & J_T & 0 & 0 & J_R \\
b_0 & a & K_3 & J_R & 0 & 0 \\
0 & 0 & b & K_4 & J_R & 0 \\
c_0 & 0 & 0 & c & K_5 & J_A \\
0 & e & 0 & 0 & d & K_6
\end{bmatrix}
$$

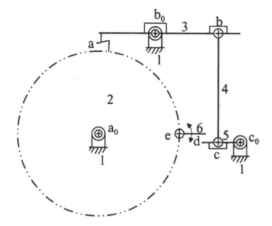

Figure 6.11 Mechanism sketch of the water wheel steelyard-clepsydra device

where the diagonal element is the member type; the upper-right nondiagonal element is the joint type; the lower-left nondiagonal element is the joint label.

The characteristics of the topological structure of this design are:
1. It is a planar six-bar mechanism with eight joints.
2. It has a ground link (member 1, K_G), a water wheel (member 2, K_2), an upper balancing lever (member 3, K_3), a connecting rod (member 4, K_4), an upper stopping tongue (member 5, K_5), and a water-receiving scoop (member 6, K_6).
3. It has one upper stopping joint (J_T), one cam joint, and six revolute joints.
4. It has one degree of freedom.
5. It has one ground link with multiple incident joints.

6.4.2 Generalized kinematic chains

Once the design specifications and topological characteristics of the water wheel steelyard-clepsydra device are obtained, the next step of the reconstruction design methodology is to obtain the atlas of generalized kinematic chains with six members and eight joints. And, there are nine (6, 8) generalized kinematic chains, as shown in Figure 4.14 and again shown in in Figure 6.12.

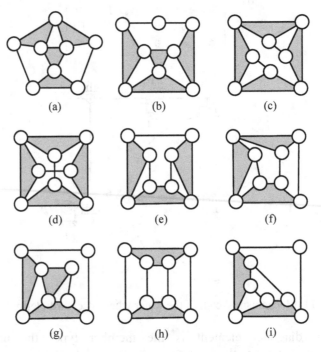

Figure 6.12 Atlas of (6, 8) generalized kinematic chains for waterwheel steelyard-clepsydra device

6.4.3 Specialized chains

Once the atlas of generalized kinematic chains is obtained, all possible specialized chains can be identified according to the following substeps subject to required design constraints:
1. For each generalized kinematic chain, identify the ground link (link 1, K_G) for all possible cases.

2. For each case obtained in step 1, identify the waterwheel (link 2, K_2).
3. For each case obtained in step 2, identify the water-receiving scoop (link 6, K_6).
4. For each case obtained in step 3, identify the upper balancing lever (link 3, K_3).
5. For each case obtained in step 4, identify the upper stopping tongue (link 5, K_5).
6. For each case obtained in step 5, identify the connecting rod (link 4, K_4).

Design constraints are determined based on the concluded topological structures of Su Song's waterwheel steelyard-clepsydra device. However, based on the result of literature research, the output link (water-receiving scoop, link 6) of the time steelyard-clepsydra device does not necessarily have to lie on the waterwheel. Consequently, the adjacency constraint of link 6 and link 2 is released, and the process of specialization is carried out with two different cases. One is that the water-receiving scoop lies on the waterwheel, i.e., link 6 lies on link2; the other is that the water-receiving scoop does not lie on the waterwheel, i.e., link 6 does not lie on a link.

Case 1: link 6 lies on link 2
The design constraints for this case are as follows:
Ground link (link K_G)
1. In each generalized kinematic chain, there must be one ground link as the frame.
2. The ground link must be a multiple link.
Driving wheel (link 2)
1. Link 2 is adjacent to the ground link with a revolute joint (J_R).
2. Link 2 is adjacent to link 3 with an upper stopping joint (J_T).
Water-receiving scoop (link 6, the output link of time steelyard-clepsydra device)
1. Link 6 is the input link of the waterwheel steelyard-clepsydra device.
2. Link 6 must be a binary link.
3. Link 6 must lie on link 2.
4. Link 6 cannot be adjacent to the ground link.
Lever mechanism (including links 3–5)
1. Link 3 is adjacent to link 2 with an upper stopping joint (J_T).
2. Link 3 or link 5 is adjacent to link 6 with a cam joint (J_A).
3. If link 5 is adjacent to link 6, then link 5 operates directly or indirectly on link 3.
4. If link 5 is not adjacent to link 6, then links 5 and 4 are mutually interchangeable.

According to the process of specialization, four feasible specialized chains are obtained as shown in Figure 6.13.

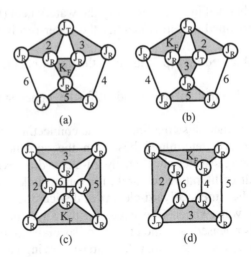

(a)   (b)

(c)   (d)

Figure 6.13 Atlas of feasible specialized chains for waterwheel steelyard-clepsydra device (link 6 on link 2)

Case 2: link 6 does not lie on link 2

The design constraints that are different from Case 1 are:
1. Link 2 must be a binary link.
2. Link 6 must not be adjacent to link 2.
3. Link 6 is adjacent to the ground link with a revolute joint (J_R).
According to the process of specialization, four feasible specialized chains are obtained as shown in Figure 6.14.

6.4.4 Reconstruction designs

In general, it is not possible to follow specific procedures to obtain new and ideal results for particularization. However, in order to improve their availability, only the motion and function requirements of mechanical devices are taken into account and the types of links and joints are kept unchanged.

Therefore, through particularizing the atlases of specialized chains shown in Figure 6.13 and Figure 6.14, a total of eight corresponding waterwheel steelyard-clepsydra mechanisms with a four-bar linkage are obtained as shown in Figure 6.15 and Figure 6.16, respectively. These four-bar linkages should have the functions of toggle effect or lever (weighing) effect. There is only one water flow circuit in the atlas of Case 1 (link 6 lies on link 2). There are two water flow circuits in the atlas of

Case 2 (link 6 does not lie on link 2). One of the water flow circuits is used to pour the water-receiving scoop of the steelyard-clepsydra. The other is used to pour the scoops of the waterwheel and is the main power source. Therefore, the water-receiving scoops of the waterwheel are fixed on the waterwheel without relative motions.

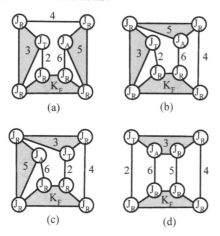

Figure 6.14 Atlas of feasible specialized chains for waterwheel steelyard-clepsydra device (link 6 not on link 2)

Including the original design shown in Figure 6.8, eight reconstruction designs of the waterwheel steelyard-clepsydra device with a four-bar linkage are obtained. Furthermore, Figure 6.17 shows the corresponding 3D solid model of the mechanisms in Figure 6.15 and Figure 6.18 shows the corresponding 3D solid model of the mechanisms in Figure 6.16. Furthermore, Figure 6.19 shows a physical reconstruction design of Su Song's waterwheel steelyard-clepsydra device.

6.5 Escapement Regulator and Modern Mechanical Clocks

Modern and ancient mechanical clocks in China and abroad are composed of five major elements: a power device, an oscillator, an escapement, a transmission mechanism, and a display device. The escapement regulator, which is made up of the oscillator and the escapement, is a critical mechanical clock technology that determines the accuracy of mechanical clocks.

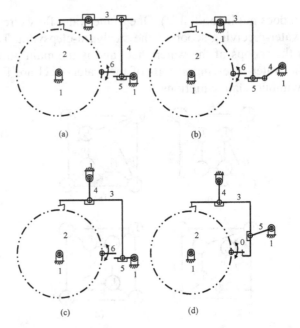

Figure 6.15   Atlas of feasible designs for the waterwheel steelyard-clepsydra device with four-bar linkage (link 6 on link 2)

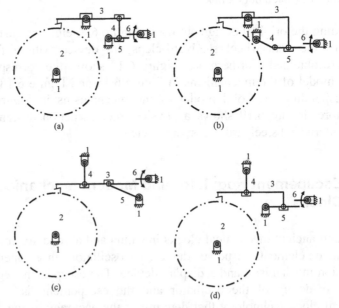

Figure 6.16   Atlas of feasible designs for the waterwheel steelyard-clepsydra device with four-bar linkage (link 6 not on link 2)

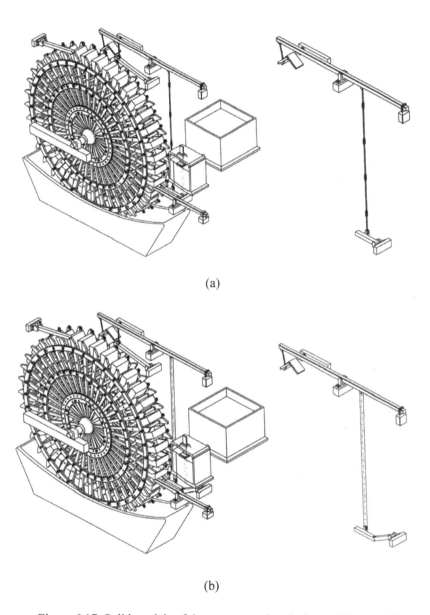

(a)

(b)

Figure 6.17 Solid models of the reconstruction designs of Figure 6.15

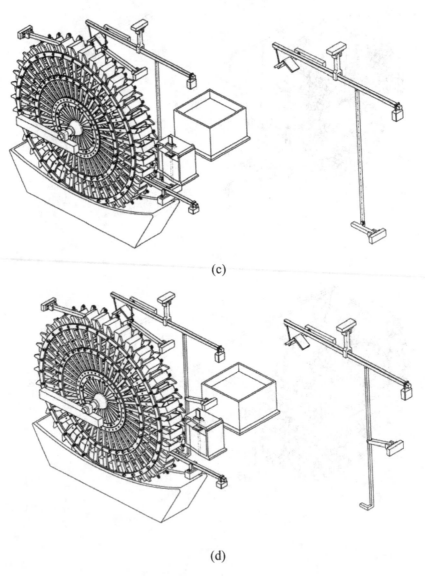

(c)

(d)

Figure 6.17 (*Continued*)

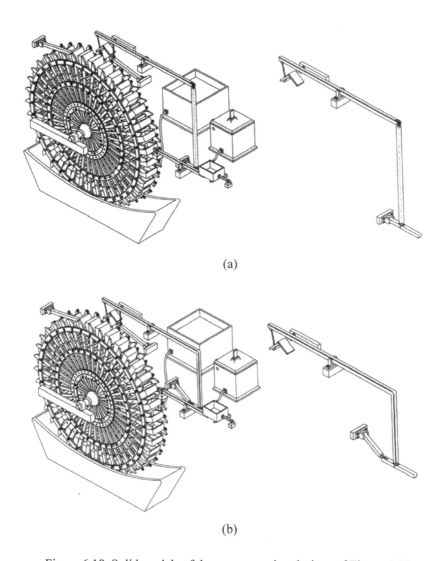

(a)

(b)

Figure 6.18 Solid models of the reconstruction designs of Figure 6.16

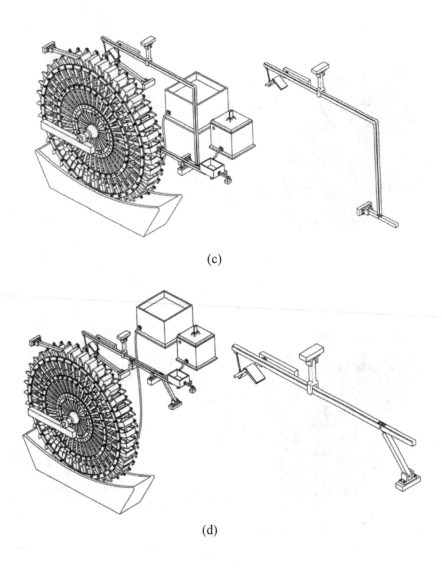

(c)

(d)

Figure 6.18 (*Continued*)

Figure 6.19 A physical reconstruction design of Figure 6.16

The purpose of the oscillator is to divide time into a stream of uniform segments. Oscillations in modern mechanical clocks and watches may be produced by either a weight or a spring [17]. In the weight system, the motive force is generated through the coaction of the centrifugal force of mass and the earth's gravity, such as in a pendulum. The weight system is used for clocks intended to be kept in one place. The spring system uses the centrifugal force of mass and the tension in springs for its continued operation, such as in a balance. Both the weight and the spring system used the escapement for the transmission of energy, unlike the time steelyard device used in Su Song's clock tower. In the time steelyard device, motive force came from the uniform water flow from the constant-level tank; the time steelyard device used the repetitive accumulation of energy and the periodic release of energy to meet the requirements of "uniformity."

Escapements are intermittent motion devices. Modern mechanical clocks and watches use different types of escapements, which convert circular motion into intermittent, back-and-forth motion. Escapements perform two major functions: to distribute energy to the oscillator to maintain movement; and to transmit driving forces to the hands, which display time. The anchor escapement, whose escape wheel is supported by the last pinion of the gear train (Figure 6.20), has the widest range of usage. On the other hand, the escape pallet and fork transforms the rotating movement into a back-and-forth motion and transmits impulses to the balance in order to produce oscillations. In return, the pallet receives the oscillation of the balance. This rhythmical order synchronizes and measures out the speed of the escape wheel by allowing only one tooth to escape at each vibration or semi-oscillation. The escapement governs the rotating speed of the escape wheel and the gear train, and prevents the mainspring from unwinding too quickly.

The water wheel linkwork escapement does not distribute energy onto the oscillator; rather, it receives impulse resulting from the periodic movement of the oscillating system. A periodic movement thus results, and the intermittent motion of the check-and-release driving wheel is used to govern the rotational speed of the driving wheel and the transmission system.

In the escapement regulator, the movement of the oscillator and the escapement should be highly synchronized to regulate the flow of energy and to enable the gear train to maintain a uniform motion. The first escapement regulator in Europe was the verge foliot escapement regulator (Figure 6.21), which was invented during the early 14th century [18]. Due to its low oscillating frequency, each oscillation resulted in error margins; thus a higher oscillation frequency was needed to increase its accuracy. By the 17th century, the verge foliot escapement regulator ultimately gave way to pendulums and the balance. Different kinds of escapements may be used with the pendulums and the balance, such as the anchor escapement and recoil escapement. However, mechanical clocks have their inherent margin of error and the development of science and technology called for more accurate and precise time-keeping instruments.

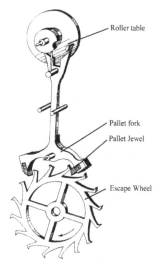

Figure 6.20 Anchor escapement [2, 18]

During the 20th century, electronic watches became very popular due to the invention of microelectronics and the knowledge of microelectronic materials. The escapement regulator was no longer the only mechanical type. By the 1950s, tuning fork electronic watches started to appear on the market. In these watches, the tuning fork oscillating frequency which tunes to the pulses coming from a driver coil was used as a basis for time keeping. The escapement was also replaced by a ratchet mechanism. During the 1960s, the watch-making industry made use of the high frequency and high stability of quartz oscillation, using electrical circuits to drive a step motor or integrated circuit with functions similar to that of an escapement. These quartz watches were accurate to within 0.1 s a day [19].

Calibration through astronomical observations is required to ensure accurate time keeping. Yet even universal time, which was derived from astronomical observations, could only attain an accuracy of 10^{-8} s. Moreover, such a degree of precision was insufficient to meet the growing demands of science and technology, especially those of military and aviation. Currently the most precise time-keeping devices are the Cs-beam atomic clocks. In this clock, cesium-133 atoms are subjected to microwave radiation, and the frequency of microwave radiation is then used as a basis to determine time. The cesium-atom clock has an accuracy of up to 10^{-13} s, equivalent to an error of about 1 s in 300,000 years. Thus, the definition of a second is no longer based on astronomical observations of the earth's revolution; rather, it is defined as "the duration of 9,192,631,772 periods of radiation that causes the cesium-133 atoms to make the transition between two closely spaced, or hyperfine, energy states." [20]. The extreme accuracy of the atomic clock is an important driver in the advancement of science and technology.

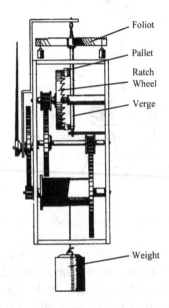

Figure 6.21 Verge foliot escapement regulator [2, 18]

6.6 Remarks

Mechanical evolution is inevitable, so the existence of every mechanism and machine has its own evolutionary history. In-depth exploration of the evolution process of scientific theories and techniques of ancient times not only yields appropriate reconstruction designs, but innovations can also be made by reviewing historical artifacts.

The reconstruction design of ancient Chinese escapement regulators is based on Su Song's water wheel steelyard-clepsydra device as an available design. It is noted that two tasks in the reconstruction design process can be very flexible. They are the establishment of design constraints and the application of mechanical evolution and variation theory. The former varies according to literature research results as well as the designers' subjective considerations, while the latter depends on the scientific theories and techniques of the subject's time period. Through an optimal integration of these two approaches, feasible reconstruction designs can be obtained.

References

1. Lin, T.Y., A systematic reconstruction design of ancient Chinese escapement regulators (in Chinese), Ph.D. dissertation, Department of Mechanical Engineering, National Cheng Kung University, Tainan, Taiwan, December 2001.
   林聰益，古中國擒縱調速器之統化復原設計，博士論文，國立成功大學機械工程學系，台南，台灣，2001 年 12 月。
2. Yan, H.S. and Lin, T.Y., "Comparison between the escapement regulators of Su Song's clock-tower and modern mechanical clocks," Proceedings of HMM2000 – the International Symposium on History of Machines and Mechanisms, Cassino, Italy, Kluwer Academic Publishers, Dordrecht, The Netherlands, pp. 141–148, 2000.
3. Yan, H.S. and Lin, T.Y., "A systematic approach to reconstruction of ancient Chinese escapement regulators," Proceedings of ASME 2002 Design Engineering Technical Conferences and Computers and Information in Engineering Conference (DETC'02), Montreal, Canada, September 2002.
4. Yan, H.S. and Lin, T.Y., "A study on ancient Chinese time laws and the time-telling system of Su Song's clock-tower," Mechanism and Machine Theory, Vol. 37, No. 1, pp. 15–33, 2002.
5. Xin Yi Xiang Fa Yao (in Chinese) by Su Song (Northern Song Dynasty), Taiwan Commercial Press, Taipei, 1969.
   《新儀象法要》；蘇頌[北宋]撰，新儀象法要，台灣商務印書館，台北，1969 年。
6. Wang, Z.D., Papers in Technical Archaeology (in Chinese), Cultural Relics Publishing House, Beijing, pp. 238–277, 1989.
   王振鐸，科技考古論叢，文物出版社，北京，第 238–277 頁，1989 年。
7. Guan, C.X., Yang, R.G., Su, K.F., Study on Su Song and Xin Yi Xiang Fa Yao (in Chinese), Jilin Literature and History Publishing Co., Zhangchun, 1991.
   管成學，楊榮垓，蘇克福，蘇頌與《新儀象法要》研究，吉林文史出版社，長春，1991 年。
8. Li, Z.C., History of Ancient Chinese Astronomical Clocks (in Chinese), China Science and Technology University Press, Hefei, 1997.
   李志超，水運儀象志，中國科技大學出版社，合肥，1997 年。
9. Hu, W.J., Xin Yi Xiang Fa Yao: A Modern Interpretation (in Chinese), Liaoning Educational Publishing House, Shenyang, 1997.

胡維佳譯註，新儀象法要，遼寧教育出版社，瀋陽，1997 年。

10. Needham, J., Wang, L., Price, D.J., Heavenly Clockwork, Cambridge University Press, London, 1960.

11. History of the Song Dynasty (New revision) (in Chinese) by Tuo Tuo (Yuan Dynasty), Ding Wen Publishing House, Taipei, Vol. 340, 1983.
    《宋史》；脫脫[元朝]等撰，新校本宋史，卷三百四十，鼎文出版社，台北，1983 年。

12. Dai, N.Z., History of Mechanics in China (in Chinese), Hebei Educational Publishing House, Shi Jia Zhuang, pp. 246–270, 1988.
    戴念祖，中國力學史，河北教育出版社，石家莊市，第 246–270 頁，1988 年。

13. Needham, J., Wang, L., and Price, D.J., "Chinese Astronomical Clockworks," Nature, Vol. 177, p. 600, 1956.

14. Shi, R.G., "Identification of Astronomical Clock and Escapement (in Chinese)," Collected Essays on History of Time-keeping Instruments, Vol. 1, First Academic Workshop on Chinese Time-keeping Instruments, Suzhou, pp. 68–75, 1994.
    施若谷，"天文鐘與擒縱器的辨析"，時計儀器史論叢，第一輯，中國計時儀器史學會，蘇州，第 68–75 頁，1994 年。

15. History of the Song Dynasty (New revision) (in Chinese) by Tuo Tuo (Yuan Dynasty), Ding Wen Publishing House, Taipei, Vol. 76, 1983.
    《宋史》；脫脫[元朝]等撰，新校本宋史，卷七十六，鼎文出版社，台北，1983 年。

16. Needham, J., Science and Civilization in China (Chinese trans.), Vol. 1, Science Press, Beijing, 1990.
    李約瑟，中國科學與文明，第 1 冊，科學出版社，北京，1990 年。

17. Zhang C.Z., (trans.), The Manufacturing of Electronic Digital Clocks and Watches (in Chinese), Xu's Foundation Publishing House, Taipei, 1968.
    張純志譯著，鐘錶製造及修理，徐氏基金會出版社，台北，1968 年。

18. Britten, F.J., Old Clocks and Watches and Their Makers, E. & F. N. Spon, Ltd., London, 1922.

19. Li, Z.C., (editor-in-chief), History of Time-keeping Instruments (in Chinese), Vol. 3, Third Academic Workshop on Chinese Time-keeping Instruments, Suzhou, pp. 8–11, 1998.
    李志超主編，時計儀器史論叢，第三輯(中國計時儀器史第三次學術研討會專輯)，中國計時儀器史學會，蘇州，第 8–11 頁，1998 年。

20. W. Wayt Gibbs (translated by Cai, Y.Z.), Ultimate Clocks (in Chinese), Scientific American, Taipei, Vol. 9, pp. 102–112, November 2002.
    吉布斯著(蔡雅芝譯)，終極時鐘，科學人雜誌，台北，第 9 期，第 102–112 頁，2002 年 11 月。

Chapter 7 South-pointing Chariots

Many ancient Chinese legends refer to the mysterious invention of the south-pointing carriage. It was a chariot on which was mounted a figure with an outstretched arm that continuously pointed south no matter which way the chariot turned. This device is called the south-pointing chariot and is found in literary records but without surviving hardware. However, various reconstructions of its designs have existed in past years.

This chapter systematically reconstructs all feasible designs of south-pointing chariots that meet the scientific and technological standards of the subject's time period [1–3]. Historical records and background of south-pointing chariots are introduced first. Then, topological structures of existing designs are analyzed and the design requirements and constraints for the structural synthesis regarding the mechanisms of south-pointing chariots are concluded. Special representations to identify different axial directions of joints and characteristics of members are defined. Finally, six design examples based on different design specifications are illustrated.

7.1 Ancient Literature

There were various literary works regarding south-pointing chariots in different dynasties in ancient China. Major historical records are as follows:
1. Gu Jin Zhu (Notes on Antiquity and Present Days), Vol. 1, Yu Fu No. 1《古今注・卷上・輿服第一》[4]

   The south pointing chariot originated from the Yellow Emperor. During the battle of Zhuolu, Chi You conjured up thick fog that blurred the vision of the Yellow Emperor's men. The Yellow Emperor thus invented a south pointing chariot to find direction, and captured Chi You.『指南車起於黃帝。與蚩尤戰於涿鹿之野，蚩尤作大霧，兵士皆迷，於是作指南車以示四方，遂擒蚩尤。』

   "Another version has it that Zhou Gong invented the south pointing chariot. When Zhou Gong ruled the land, Shang of the land of Yue sent presents. The emissary could not find his way back; thus, Zhou Gong gave him five chariots that pointed south."『舊說周公所作，周公致

太平，越裳氏重譯來獻，使者迷其歸路，周公賜軿車五乘，皆爲司南之製。』

2. Zhi Lin 《志林》[5]

    The Yellow Emperor battled with Chi You at Zhuolu. Chi You conjured up three days of fog, and the soldiers lost their way. The Yellow Emperor thus ordered his official Feng Hou to build a south pointing chariot to identify the four directions.『黃帝與蚩尤戰於涿鹿之野，蚩尤作大霧彌三日，人皆惑，帝命風后法斗機，作指南車以別四方。』

3. Huang Di Nei Zhuan 《黃帝內傳》[6]

    The Goddess Xuan built a south pointing chariot and a hodometer for the Emperor, and put them at the front and back of the vehicle, respectively.『玄女爲帝製司南車當其前，記里鼓車當其後。』

Many later literary works quoted Gu Jin Zhu (Notes on the Antiquity and Present Days) 《古今注》[4] regarding south-pointing chariots, such as:

4. Summary of Zi Zhi Tong Jian (Comprehensive Mirror for Aid in Government), Chapter 15 《資治通鑑綱目·卷十五》[7]

    The Yellow Emperor battled with Chi You at Zhuolu. Chi You conjured up thick fog, and the soldiers lost their sense of direction. The Emperor thus ordered the building of a south pointing chariot. During the reign of Zhou Cheng Wang, Shang of the land of Yue sent an emissary to bring presents. The emissary could not find his way home. Zhou Gong thus gave him a chariot that pointed south.『黃帝與蚩尤戰於涿鹿，蚩尤起大霧，將士不知所之，帝遂作指南車。周成王時越裳氏重譯來獻，使者迷失歸路，周公賜軿車以指南。』

5. Tai Ping Yu Lan, Chapter 775, Chariots 4 《太平御覽·卷七七五·車部四》[8]

    The south pointing chariot originated from the Yellow Emperor.『指南車起於黃帝』

    Gui Gu Zi said, 'Shen of the land of Su gave Zhou Wen Wang white pheasant as present. For fear that Shen might become lost on his way home, Zhou Gong built a south pointing chariot for Shen.『鬼谷子云：「肅慎氏獻白雉於文王，還恐迷路，周公因作指南車以送之。」』

6. Song Shu, Li Zhi (Annals of Rites) 《宋書·禮志》[9]

    The south pointing chariot was first built by Zhou Gong, who gave it to an emissary from a land far away. When one was lost in direction, the device always pointed to the south.『指南車始於周公所作，以送荒

外遠使。地域平漫，迷於東西，造立此車，使常知南北也。』

The existence of south-pointing chariots in the Han Dynasty (206 BC–AD 220) can be inferred from the following references:

7. Xi Jing Za Ji《西京雜記》[10]

With a south pointing chariot, one could drive the four-horse chariot and stay on course. 『司南車，駕四，中道。』

8. Song Shu, Li Zhi (Annals of Rites)《宋書・禮志》[9]

Gui Gu Zi said, 'When people from the land of Zheng mined for jade, they brought a south pointing chariot along to find directions.' The chariot disappeared during the Chin and Han dynasties. Zhang Heng of the Late Han period recreated the device, but it was lost during the upheaval at the end of the Han Dynasty.「鬼谷子云：『鄭人取玉，必載司南，爲其不惑也。』至于秦漢，其至無聞。後漢張衡，始復創造，漢末喪亂，其器不存。」

9. Wu Du Fu《吳都賦》[11]

A south pointing chariot was placed in front to point direction. The wheels of the chariot and the chains made noises.『俞騎騁路，指南司方，出車檻檻，被鍊鏘鏘。』

10. Wei Lue《魏略》[12]

[O]rdered the court academician Ma Jun to build a south pointing chariot ...『…使博士馬鈞造指南車 …』

11. History of the Three Kingdoms, Wei Shu, Chapter 29《三國志・魏書・卷二十九》[13]

At the time, there was Ma Jun of Fu Fong, who was very skillful. ... Ma Jun engaged in a heated discussion on the south pointing chariot with his aide and general. The two did not believe the device had existed. Ma Jun said, 'It existed. It is the want of thought about it, and therefore people have forgotten it.' The two jeered and said, 'Your personal name is Jun and your courtesy name is De Heng. Jun is the mold of a device, while Heng is used to measure weight. If there is no standard for measuring weight, how can a mold be built?' Ma Jun said, 'Arguments are useless. We can try to build one to end all arguments.' The two requested the Emperor to ask Ma Jun to build a south pointing chariot. Upon its completion, the device was unusual and staggering, and people became convinced of Ma's skill.『時有扶風馬鈞，巧思絕世 ... 先生爲給事中，與常侍高堂隆、驍騎將軍秦朗爭論於朝，言及指南車，二子謂古無指南車，記言之虛也，先生曰：「古有之，未之思耳，夫何遠之有！」二子哂之曰：「先生名鈞字德衡，鈞者器之模也，而衡者所以定物之輕重；輕重無準

而莫不模哉！」先生曰：「虛爭空言，不如試之易效也。」於是
二子遂以白明帝，詔先生作之，而指南車成，此一異也，又不可
以言者也，從是天下服其巧矣。』

12. Song Shu, Li Zhi (Annals of Rites)《宋書·禮志》[9]
*"Gao Tang-long and Qin Lang of Wei, both learned individuals, were
arguing in court that the south pointing chariot did not exist and
historians were mistaken. The Emperor of Wei ordered Ma Jun to
build the device. When the device was completed, the Wei kingdom
collapsed and was replaced by that of Jin.* 『魏高堂隆、秦朗皆博聞
之士，爭論於朝云：無指南車，記者虛說。明帝青龍中令馬鈞更
造之，而車成晉亂復亡。』

13. Jin Shu, Zhi, Chapter 25 and Zhi Chapter 15, Yu Fu《晉書·志·卷二
十五；志·第十五·輿服》[14]
*The south pointing chariot was lost when crossing the river. It was
only during the fifth year of Yi Xi when Liu Yu invaded Guang Gu that
the device was re-discovered. Liu ordered craftsman Zhang Gang to
repair the device. On the 13th year, Liu occupied Guan Zhong and
obtained another south pointing chariot and a hodometer. The
implements were now complete.* 『指南車，過江亡失，及義熙五
年，劉裕屠廣固，始復獲焉，乃使工人張綱補緝周用。十三年，
裕定關中，又獲司南、記里諸車，制度始備。』

14. Nan Qi Shu, Lie Zhuan, Chapter 34 and Lie Zhuan, No. 15, Liu Xiu
《南齊書·列傳·卷三十四；列傳·第十五·劉休》[15]
*During the late Song Dynasty, the Emperor ordered the building of a
south pointing chariot based on Liu Xin's idea. Liu Xiu and Wang
Seng-qian were ordered to test the device.* 『宋末，上造指南車，以
休有思理，使與王僧虔對共監試。』

15. Nan Qi Shu, Lie Zhuan, Chapter 52 and Lie Zhuan, No. 33, Wen Xue,
Zu Chong-zhi《南齊書·列傳·卷五十二；列傳·第三十三·文學·
祖沖之》[15]
*In the beginning, when Emperor Wu of the Song Dynasty occupied
Guanzhong, he obtained the south pointing chariot of Yao Xing of the
kingdom of Chin. The device had an outer shell but no inner mecha-
nism; therefore, it had to be set in motion by a man hidden inside the
device in every trip. During the middle Sheng Ming years of the Song
Dynasty, Emperor Tai Zu aided state administration and ordered Zu
Chong-zhi to repair the device according to the ancient method. Tzu
Chong-zhi repaired the copper device and made its mechanism rotate.
It indicated directions perfectly, that had not been possible after Ma Jun.*

At the time, a stallion trainer from the north claimed that he could also build a south pointing chariot. Emperor Tai Zu allowed Zu Chong-zhi and the trainer built their own version of the device, and tested the devices at the Le You Park. The device by the stallion trainer did not function well and it was burned to destroy. 『初，宋武帝平關中，得姚興指南車，有外型而無機巧，每行，使人內轉之。昇明中，太祖輔政，使沖之追修古法。沖之改造銅機，圓轉不窮，而司方如一，馬鈞以來未有也。時有北人索馭驎者，亦云能造指南車，太祖使與沖之各造，使於樂遊苑對共校試，而頗有差僻，乃毀焚之。』

16. Nan Qi Shu, Lie Zhuan, Chapter 59 and Lie Zhuan, No. 40, Rui Rui Lu《南齊書・列傳・卷五十九；列傳・第四十・芮芮虜》[15]
    When the king of Rui Rui asked for medicines and implements, Emperor Shi Zu's imperial edict said, 'It must be understood that medicines, textiles, south pointing chariots, time pieces, etc., are present, but the artisans who built them are no longer around.' 『芮芮王求醫工等物，世祖詔報曰：「知須醫及織成錦工、指南車、漏刻，… 此雖有其器，工匠久不復存。」』

17. Nan Shi, Ben Ji, Chapter 1 and Song Ben Ji, Part 1, No. 1, Wu Di《南史・本紀・卷一；宋本紀・上・第一・武帝》[16]
    City Changan was prosperous and wealthy. The emperor collected sacrificial vessels, the armillary sphere, sun dial, hodometer, south pointing chariot, and jade seal of the Qin emperor, and delivered them back to his capital. 『長安豐稔，帑藏盈積，帝先收其彝器、渾儀、土圭、記里鼓、指南車、及秦始皇玉璽送之都。』

18. Nan Shi, Lie Zhuan, Chapter 47 and Lie Zhuan, No. 37, Liu Xiu《南史・列傳・卷四十七；列傳・第三十七・劉休》[16]
    At the end of the Song Dynasty, Liu Xiu built a south pointing chariot. The emperor ordered him to test it with Wang Seng-qian. 『宋末，造指南車，高帝以休有思理，使與王僧虔對共監試。』

19. Jiu Tang Shu, Ben Ji, Chapter 15 and Ben Ji, No. 15, Xian Zong Last Part, Yuan He Year 10《舊唐書・本紀・卷十五；本紀・第十五・憲宗下・元和十年》[17]
    In the Geng Shen year, the new south pointing chariot and the hodometer were built. Seventy-two palace officials were sent to place the chariot at the city temple. 『庚申，新造指南車、記里鼓，出宮人七十二人置京城寺觀 …』

20. Jiu Tang Shu, Ben Ji, Chapter 16 and Ben Ji, No. 16, Mu Zong Li Heng, before Chang Qing Year 1《舊唐書・本紀・卷十六；本紀・第

十六・穆宗李恆・長慶元年以前》[17]

The Cheng De army governor Wang Cheng-zong died. His younger brother Cheng-yuan wrote the palace asking for military commission. The palace sent Bo Qi, an attendant of the emperor, to console him. In the Xin Si year, Jin Gong-liang successfully repaired a south pointing chariot and a hodometer. 『成德軍節度使王承宗卒，其弟承元上表請朝廷命帥，遣起居舍人柏耆宣慰之。辛巳，金公亮修成指南車、記里鼓車。』

21. Song Shi, Ben Ji, Chapter 9 and Ben Ji, No. 9, Ren Zong Zhao Zhen One, Tian Sheng Year 5《宋史・本紀・卷九；本紀・第九・仁宗趙禎一・天聖五年》[18]

In the Geng Zi year, an emissary was sent to He Bei to understand and appease. In the Jen Yin year, a south pointing chariot was built again. In the Xin Hai year, banquet was offered in the morning at the Jing Ling Palace. 『庚子，遣使河北體量安撫。壬寅，復作指南車。辛亥，朝饗景靈宮。』

22. Song Shu, Zhi, Chapter 18 and Zhi, No. 8, Li Five《宋書・志・卷十八；志・第八・禮五》[9]

Shi Hu ordered Xie Fei and Yao Xing ordered Ling Hu-hseng to rebuild it. In the 13th year of the Yi Hsi period of Emperor An of the Chin Dynasty, Song Emperor Wu conquered Changan and obtained the chariot. The chariot was built like a hodometer. A wooden man was placed on the chariot that pointed south. The chariot may move in any direction, but the wooden man would always point to the south. It was placed in front of an imperial convoy. The chariot was built by Rong Di people, and the mechanism was not refined. Although the mechanism was supposed to point to the south, it was inaccurate. Its movements had to be manually corrected. Zu Chong-zhi of Fanyang had good ideas and often said that he would rebuild the chariot. During the latter Sheng Ming period when Qi Wang served as prime minister, Emperor Shun ordered Zu Chong-zhi to build the chariot. When it was finished, it was tested by army motivator and administrator of the state capital, Wang Heng-qian, and imperial emissary and minister Liu Xiu. They found that the chariot was refined, and no adjustments were necessary even after hundreds of rotation. The chariot was also referred to as the south pointing vessel during the Chin Dynasty. A Suo Lu local, Tuo Ba-tao, ordered his worker, Guo Shan-ming, to build the south pointing chariot, but it was not finished for a long time. Ma Yue of Fu Fong tried to build one but was killed by San Ming when his work neared completion. 『... 石虎使解飛，姚興使令狐生又造

焉。安帝義熙十三年，宋武帝平長安，始得此車。其制如鼓車，設木人於車上，舉手指南。車雖回轉，所指不移。大駕鹵簿，最先啓行，此車戎狄所制，機數不精，雖曰指南，多不審正。回曲步驟，猶須人功正之。范陽人祖沖之，有巧思，常謂宜更構造。宋順帝昇明末，齊王爲相，命造之焉。車成，使撫軍丹陽尹王僧虔、御使中丞劉休試之。其制甚精，百曲千回，未常移變，晉代又有指南舟。索虜拓跋燾使工人郭善明造指南車，彌年不就。扶風人馬岳又造，垂成，善明酖殺之。』

23. Wei Shu, Li Zhi《魏書·禮志》[19]

During the second year of the Tian Xing period, the Wei emperor ordered the rites officer to collect old records to create protocols for the imperial convoys. ... One version has it that the convoy was composed of thousands of chariots and drivers, where the battle chariots were placed in front while the soldiers marched alongside and behind shoulder-to-shoulder and leading out. The vanguards rode on animal skin cover chariots, their weapons were concealed. The yun han chariot had a south pointing device on and a leopard's tail hanging on its back section.『太祖天興二年，命禮官挼採古事。制三駕鹵簿，一日大駕 ... 千乘萬騎、魚麗鴈行，前驅皮軒，闟戟芝蓋，雲罕指南復殿豹尾。』

24. Jin Shu, Zhi, Chapter 25 and Zhi Chapter 15, Yu Fu《晉書·志·卷二十五；志·第十五·輿服》[14]

The south pointing chariot was drawn by four horses. It was three-story high and the four corners were decorated with golden dragons and bird's feathers. A wooden human figure with colorful garb stood on the chariot. Regardless of which direction the chariot was facing the hand of the figure always pointed south. It was placed in front of the imperial convoy.『司南車一名指南車，駕四馬。其制如樓三級，四角金龍銜羽葆。刻木爲仙人，衣羽衣，立車上，車雖回轉，而手常南指。大駕出行，爲先啓之乘。』

25. Nan Qi Shu, Zhi, Chapter 17 and Zhi, No. 9, Yu Fu《南齊書·志·卷十七；志·第九·輿服》[15]

The south pointing chariot was a box with walls adorned with heavenly cloth. Inside the box, there were dragons and peacock feathers decorated on the four corners. Black cloth was used as layered curtains and the wheels were painted with designs. The chariot was pulled by a cow and was adorned with copper.『指南車，四周廂上施屋，指南車衣裙襦天衣，在廂中。上四角皆施龍子竿，縣雜色真孔雀毦毛，烏布皁複幔，漆畫輪，駕牛，皆銅校飾。』

26. Shi Shu, Zhi, Chapter 10 and Zhi, No. 5, Li Yi Five《隋書·志·卷十；志·第五·禮儀五》[20]

The south pointing chariot was placed in front of the convoy every time. During the early Han Dynasty, a yu er rode in the chariot that led the convoy. Officer Zuo Tai Fu said 'The south pointing chariot was placed in front to point direction with a yu er riding on it.' Later, the driver was discarded and only the chariot remained.『指南車，大駕出，爲先啓之乘。漢初，置俞兒騎，並爲先驅。左太傅曰：「俞騎騁路，指南司方。」後廢其騎而存其車。』

27. Jiu Tang Shu, Zhi, Chapter 45 and Zhi, No. 25, Yu Fu, Che Yu, Tian Zi Che Yu 《舊唐書·志·卷四十五；志·第二十五·輿服·車輿·天子車輿》[17]

Protocols of Tang Dynasty. The five imperial chariots were jade, gold, elephant skin, animal skin, and wood. There were also the geng gen chariot, an chariot, and si wang chariot, a total of eight chariots for conveying people. There were likewise the south pointing chariot, hodometer, white heron chariot, luan qi chariot, pi er chariot, xuan chariot, leopard tail chariot, ram chariot, and yellow battle ax chariot. The leopard tail and yellow battle ax chariots were not included in the convoy during the Wu De period. They were added during the Zhen Guan period. The yellow battle ax was changed to golden battle ax according to protocols during the Tian Bao period. There were twelve chariots for retinues and honor guards were placed before and after the emperor's chariot in the convoy.『… 唐制，天子車輿有玉輅、金輅、象輅、革輅、木輅，是爲五輅，耕根車、安車、四望車，已上八等，並供服乘之用。其外有指南車、記里鼓車、白鷺車、鸞旗車、辟惡車、軒車、豹尾車、羊車、黃鉞車，豹尾、黃鉞二車，武德中無，自貞觀已後加焉。其黃鉞，天寶元年制改爲金鉞。屬車十二乘，並爲儀仗之用。大駕行幸，則分前後，施於鹵簿之內。』

28. New Revised Jiu Tang Shu, Zhi, Chapter 44 and Zhi, No. 24, Zhi Guan Three, Tai Pu Si《新校本舊唐書·志·卷四十四；志·第二十四·職官三·太僕寺》[21]

Cheng Huang Shu: one ling official, from the 7th rank down; one cheng official, from the 8th rank down; one fu official; two shi officials, eight dian shi officials; 140 drivers; an official of goat-driven cart, and six zhang gu officials. All these were responsible for handling the emperor's chariots, inventory, and animal training. Their secondary tasks were to keep records of events, implements, chariots, and uniforms for every five chariots ridden. There were twelve minor chariots:

south pointing chariot, hodometer, white heron chariot, luan qi chariot, pi er chariot, animal skin chariot, geng gen chariot, an chariot, si wang chariot, ram chariot, yellow battle ax chariot, leopard tail chariot, that decorations were recorded in Yu Fu Zhi. 『 … 乘黃署：令一人，從七品下。丞一人，從八品下。府一人，史二人，典事八人，駕士一百四十人，羊車小吏十四人，掌固六人。令掌天子車輅，辨其名數與馴馭之法。丞爲之貳。凡乘輿五輅，事具輿服志也。皆有副車，又有十二車，曰指南車、曰記里鼓車、白鷺車、鸞旗車、辟惡車、皮軒車、耕根車、安車、四望車、羊車、黃鉞車、豹尾車，其車飾見輿服志也。』

29. New Tang Shu, Zhi, Chapter 23, Vol. 1 and Zhi, No. 13, Vol. 1, Jia 《新唐書・志・卷二十三・上；志・第十三・上・駕》[22]

[F]ollowed by the zhu que group. Then, followed by the south pointing chariot, hodometer, white heron chariot, pi e chariot, and animal skin chariot. There were one navigator and fourteen drivers. … 『 … 次朱雀隊。次指南車、記里鼓車、白鷺車、鸞旗車、辟惡車、皮軒車，皆四馬，有正道匠一人，駕士十四人，…』

30. New Tang Shu, Zhi, Chapter 23, Vol. 3 and Zhi, No. 13, Vol. 3, Da Jia Lu Bu Gu Chui《新唐書・志・卷二十三・下；志・第十三・下・大駕鹵簿鼓吹》[22]

For the xiao jia, the yu shi official and the numbers of south pointing chariot, hodometer, luan qi chariot, animal skin chariot, elephant-skin mu san lu, geng gen chariot, ram-driven chariot, yellow battle ax chariot, leopard tail chariot, rhino chariot, small push cart and chariots, and bands were reduced to half as compared with the da jia. 『 … 小駕，又減御史大夫、指南車、記里鼓車、鸞旗車、皮軒車、象革木三路、耕根車、羊車、黃鉞車、豹尾車、屬車、小輦、小輿，諸隊及鼓吹減大駕之半。』

31. New Tang Shu, Zhi, Chapter 24 and Zhi, No. 14, Che Fu, Tian Zi Zhi Che《新唐書・志・卷二十四；志・第十四・車服・天子之車》[22]

There were also ten chariots of retinues: south pointing chariot, hodometer, white heron chariot, luan gi chariot, pi er chariot, animal skin chariot, ram chariot, geng gen chariot, si wang chariot, and an chariot. The chariots in an imperial convoy were placed at the front and back; whereas in large gatherings, they were placed on both sides. 『又有屬車十乘：一曰指南車，二曰記里鼓車，三曰白鷺車，四曰鸞旗車，五曰辟惡車，六曰皮軒車，七曰羊車，與耕根車、四望車、安車爲十乘。行幸陳於鹵簿，則分前後；大朝會，則分左右。』

32. Xuan He Lu Bu Ji《宣和鹵簿記》[23]

In early Tang Dynasty, the term south pointing chariot was known but the chariot was already destroyed. Yang Wu-lian, a skillful builder, was ordered to repair the chariot but was unsuccessful. Many chariots were built during the middle Kai Yuan period. During the middle Yuan He period, craftsman Jin Zhong-yi built a south pointing chariot and a hodometer as Emperor Xian Zong watched from the Lin De Palace. 『唐初指南車，有其名而車破壞。將作大將楊務廉性巧，奉勑改作，終不能至。開元中衛普普作車，令直少府監。元和中巧工金忠義(公立)作指南車，記里鼓，憲宗於麟德殿觀之。』

33. Yu Hai, Che Fu Bu《玉海‧車服部》[23]

During the tenth year in the Yuan He period of Emperor Xian Zong, a new south pointing chariot was built and inspected in the Lin De Palace. 『憲宗元和十年閱新作指南車於麟德殿。』

 On the 10th lunar month of the 15th year in the Yuan He period, Jin Gong-liang restored the south pointing chariot and the hodometer. 『元和十五年十月辛巳金公亮修成指南記里鼓車。』

34. Song Shi, Zhi, Chapter 148 and Zhi, No. 101, Lu Bu Yi Fu《宋史‧志‧卷一百四十八；志‧第一百一‧鹵簿儀服》[18]

The driver of the geng gen chariot used the phoenix and standing grain as design, the jin xian chariot used auspicious rui lin, the ming yuan chariot used paired phoenixes, the ram chariot use auspicious ram, the south pointing chariot used peacock, the hodometer and yellow battle ax chariots used paired geese, the white heron chariot used soaring heron, the luan qi chariot used the auspicious luan bird, the chong de chariot used pi xie, the animal skin chariot used the tiger, the shu chariot used the crane, and the leopard tail chariot used the standing leopard. 『耕根車駕士以鳳銜嘉禾，進賢車以瑞麟，明遠車以對鳳，羊車以瑞羊，指南車以孔雀，記里鼓、黃鉞車以對鵝，白鷺車以翔鷺，鸞旗車以瑞鸞，崇德車以辟邪，皮軒車以虎，屬車以雲鶴，豹尾車以立豹。』

35. Song Shi, Zhi, Chapter 149 and Zhi, No. 102《宋史‧志‧卷一百四十九；志‧第一百二》[18]

Yu Fu one, five lu chariots, big lu chariot, big push cart, fang-ting-nian chariot, phoenix nian chariot, xiao-yao-nian chariot, seven treasures nian chariot, small yu chariot, yao-yu chariot, geng-gen chariot, jin-xian chariot, ming-yuan chariot, ram chariot, south pointing chariot, hodometer chariot, white heron chariot, luan-qi chariot, chong-de chariot, animal skin chariot, yellow battle ax chariot, leopard tail chariot, shu chariot, five chariot, liang chariot, xiang-feng-wu-yu chariot, xing-lou-yu chariot, twelve divinities yu chariot, percussion instrument and drum chariot, zhong-gu-lou-yu chariot …『… 輿服一，五輅，大輅，大輦，芳亭輦，鳳輦，逍遙輦，平輦，七寶輦，小輿，腰輿，耕根車，進賢車，明遠車，羊車，指南車，記里鼓車，白鷺車，鸞旗車，崇德車，皮軒車，黃鉞車，豹尾車，屬車，五車，涼車，相風烏輿，行漏輿，十二神輿，鉦鼓輿，鍾鼓樓輿 …』

36. Song Shi, Zhi, Chapter 149 and Zhi, No. 102, Yu Fu One, South Pointing Chariot《宋史·志·卷一百四十九；志·第一百二·輿服一·指南車》[18]

The south pointing chariot was reddish. Both boxes were painted with green dragon and white tiger, and flowers, birds, entertainers, arch, and perfume pouches on the four sides. There was a figure on top that pointed south whichever direction the chariot traveled. Four horses were secured to one shaft that had a phoenix tip. There were originally eighteen drivers, but they were increased to thirty during the Yong-xi period of Emperor Tai Zhong. During the fifth year of the Tian Sheng period of Emperor Ren Zong, engineering officer Yan Su began to build a south pointing chariot. Yan Su told the emperor: 'When the Yellow Emperor battled with Chi You in Zhuolu, Chi You conjured up thick fog and the soldiers lost their sense of direction. The emperor thus ordered the building of a south pointing chariot. And, during the reign of Zhou Cheng Wang, Shang of the land of Yue sent an emissary to bring presents. The emissary could not find his way home. Zhou Gong thus gave him a south pointing chariot. After that, the method was lost. Zhang Heng of Han Dynasty and Ma Jun of Wei period built it, but the device was lost due to turmoil. When Emperor Wu of the Song Dynasty conquered the city Changan, he tried to build the device but was unsuccessful. Zu Chong-zhi also tried to replicate the device. During the later Wei period, Emperor Tai Wu ordered Guo Shan-ming to build the device, but Guo Shan-ming tried for many years and was unsuccessful. Ma Yue of Fu Feng was commissioned to build one, but was killed by Guo Shan-ming before the device was completed. The method of building the device was thus lost again. During the middle years of the Yuan He period of the Tang Dynasty, Jin Gong-li brought a south pointing chariot and a hodometer to the emperor, and Emperor Xian Zong inspected them in Lin De Palace in preparation for the imperial convoy. From the Five Dynasties until the current administration, the method had not been known. Today, the chariot can be built using the following method. There is a single-shaft in the chariot. There is an extra construction on the outer chassis of the chariot.

The wooden figure stands on top and raises its hand to point to the south. There are nine gears with a total of 120 teeth. There are two base gears six chi in height and their circumference is nineteen chi. On them are two smaller gears with two chi four cun in diameter and a circumference of seven chi and two cun. There are 24 teeth on them, and the distance between teeth is three cun. Two small gears with thre cun in diameter are attached to the end of the shaft and connected by an axle. The small left and right level gears are one chi and two cun in diameter with 12 teeth. The big center level gear is four chi and eight cun in diameter with a circumference of fifteen chi. It has 48 teeth with a distance of three cun between teeth. There is an axle at the center of the chariot. It is eight chi high and three cun in diameter. On top of the axle is the wooden figure that points to the south. If the chariot moves to the east, the shaft rotates to the right. The base and smaller gears on the right advance 12 teeth, that connect with the small level gear at the right one full circle. This in turn connects with the big level gear at the center causing it to turn one-fourth circle to the left, turning 12 teeth. Therefore, if the chariot moves to the east, the wooden figure points to the south. If the chariot moves to the west, the shaft rotates to the left. The base and smaller gears on the left advance 12 teeth, that connect with the small level gear at the left one full circle. This in turn connects with the big level gear at the center causing it to turn one-fourth circle to the right, turning 12 teeth. Therefore, if the chariot moves to the west, the wooden figure points to the south. If the chariot moves to the west, the shaft rotates to the left. If the chariot moves to the north, east, west, the mechanics are the same. The emperor can issue an edict to build a south pointing chariot this way.『指南車，一曰司南車。赤質，兩箱畫青龍、白虎，四面畫花鳥，重臺，勾闌，鏤拱，四角垂香囊。上有仙人，車雖轉而手常南指。一轅，鳳首，駕四馬。駕士舊十八人，太宗雍熙四年，增爲三十人。仁宗天聖五年，工部郎中燕肅始造指南車。肅上奏曰：『黃帝與蚩尤戰于涿鹿之野，蚩尤起大霧，軍士不知所向，帝遂作指南車。周成王時，越裳氏重譯來獻，使者惑失道，周公賜軿車以指南。其後，法俱亡。漢張衡、魏馬鈞繼作之，屬世亂離，其器不存。宋武帝平長安，嘗爲此車，而制不精。祖沖之亦復造之。後魏太武帝使郭善明造，彌年不就，命扶風馬岳造，垂成而爲善明鴆死，其法遂絕。唐元和中，典作官金公立以其車及記里鼓上之，憲宗閱於麟德殿，以備法駕。歷五代至國朝，不聞得其制者，今創意成之。其法：用獨轅車，車箱外籠上有重構，立木仙人於上，引臂南指。用大小輪九，合齒一百二十。足輪二，高六尺，

圍一丈八尺。附足立子輪二，徑二尺四寸，圍七尺二寸，出齒各
二十四，齒間相去三寸。轅端橫木下立小輪二，其徑三寸，鐵軸
貫之。左小平輪一，其徑一尺二寸，出齒十二；右小平輪一，其
徑一尺二寸，出齒十二。中心大平輪一，其徑四尺八寸，圍一丈
四尺四寸，出齒四十八，齒間相去三寸。中立貫心軸一，高八
尺，徑三寸。上刻木爲仙人，其車行，木人指南。若折而東，推
轅右旋，附右足子輪順轉十二齒，擊右小平輪一匝，觸中心大平
輪左旋四分之一，轉十二齒，車東行，木人交而南指。若折而
西，推轅左旋，附左足子輪隨輪順轉十二齒，擊左小平輪一匝，
觸中心大平輪右轉四分之一，轉十二齒，車正西行，木人交而南
指。若欲北行，或東，或西，轉亦如之。詔以其法下有司製
之。』」

37. Song Shi, Zhi, Chapter 149 and Zhi, No. 102, Yu Fu One, South-
Pointing Chariot《宋史‧志‧卷一百四十九；志‧第一百二‧輿服一
‧指南車》[18]

*During the first year of the Da Guan period, Wu De-ren of the court
attendant office also presented plans for building a south pointing
chariot and a hodometer. When the two chariots were built, they were
used during the sacrificial rites to the ancestor that year. The south
pointing chariot was one zhang, one chi, one cun and five fen (分, ancient Chinese length of one tenth of a cun) long, nine and a half chi
wide, and one zhang and nine cun deep. The diameter of the gear was
five chi and seven cun, and the shaft was one zhang and five cun.
There were two layers on the upper and lower portion of the chariot
box. A screen was installed at the middle. There was a human figure
holding a cane, a turtle and crane by its side, and four child figures
holding tassels on the four corners. A roof was installed on top. There
were 13 horizontal gears each with one chi, eight cun and five fen diameter and circumference of five chi, five cun and five fen. The gears
had 32 teeth each, and the distance between the teeth was one cun and
eight fen. The central axle passed through the screen. There were 13
mid-sized to larger level gears below. The diameter of the gear was
three chi and eight cun; the circumference was one zhang, one chi and
four cun. The gears had 100 teeth, and distance between teeth was one
cun, two fen and five li (釐, ancient Chinese length of one tenth of a
fen). The gears were connected to the ones above and moved horizon-
tally and vertically. Two small level gears each had a hanging weight.
The gears had a diameter of one chi and one cun, a circumference of
three chi and three cun. They each had 17 teeth, and the distance between teeth was one cun and nine fen. There were secondary gears on*

both their sides. The diameter of these gears was one chi and five cun, the circumference was four chi, six cun and five fen. Each gear had 24 teeth, and the distance between teeth was two cun and one fen. There were also two auxiliary gears on each side. The diameter of the lower auxiliary gears was two chi and one cun; circumference was six chi and three cun. There were 32 teeth, and distance between teeth was two cun and one fen. Diameter of the upper auxiliary gears was one chi and two cun; circumference was three chi and six cun. There were 32 teeth, and distance between teeth was one cun and one fen. There was one vertical gear on each side of the chariot shaft. The diameter of the gear was two chi and two cun, and circumference was six chi and six cun. The gears had 32 teeth, and the distance between teeth was two cun, two fen and five li. There were also small wheels at the back of the chariot shaft. The wheels had no teeth, and were tied to the bamboo where the rope was tied to rods of both sides.『大觀元年，內侍省吳德仁又獻指南車、記里鼓車之制，二車成，其年宗祀大禮始用之。其指南車身一丈一尺一寸五分，闊九尺五寸，深一丈九寸，車輪直徑五尺七寸，車轅一丈五寸。車箱上下爲兩層，中設屏風，上安仙人一執杖，左右龜鶴各一，童子四各執纓立四角，上設關戾。臥輪一十三，各徑一尺八寸五分，圍五尺五寸五分，出齒三十二，齒間相去一寸八分。中心輪軸隨屏風貫下，下有輪一十三，中至大平輪。其輪徑三尺八寸，圍一丈一尺四寸，出齒一百，齒間相去一寸二分五釐，通上左右起落。二小平輪，各有鐵墜子一，皆徑一尺一寸，圍三尺三寸，出齒一十七，齒間相去一寸九分。又左右附輪各一，徑一尺五寸五分，圍四尺六寸五分，出齒二十四，齒間相去二寸一分。左右疊輪各二，下輪各徑二尺一寸，圍六尺三寸，出齒三十二，齒間相去二寸一分；上輪各徑一尺二寸，圍三尺六寸，出齒三十二，齒間相去一寸一分。左右車轅上各立輪一，徑二尺二寸，圍六尺六寸，出齒三十二，齒間相去二寸二分五釐。左右後轅各小輪一，無齒，繫竹並索在左右軸上，遇右轉使插羽，鏧纓，攀胸鈴拂，緋絹扉，錦包尾。』

38. Jin Shi, Zhi, Chapter 41 and Zhi, No. 22, Yi We《金史·志·卷四十一；志·第二十二·儀衞》[24]
The coming of the heavenly family ... south pointing chariot, hodometer, thirty persons each.『天眷法駕 … 指南車，記里鼓車，各三十人 …』
The imperial convoy ... third year of the Da Ding period of the Emperor Shi Zong ... south pointing chariot and hodometer, both need

twelve persons. 『大駕鹵簿，世宗大定三年 … 指南、記里鼓車，皆十二人。』

39. Jin Shi, Zhi, Chapter 43 and Zhi, No. 24, Yu Fu Vol. 1, Imperial Chariot《金史・志・卷四十三；志・第二十四・輿服上・天子車輅》[24]

In the 11th year of the Da Ding period, there was to be an event in Nan Jiao. The office of temple protocols was ordered to inspect details of the event. The imperial convoy was to have jade, gold, elephant skin, animal skin, and wood chariots, one geng gen chariot, one ming yuan chariot, one south pointing chariot, one hodometer, one chong de chariot, one animal skin chariot, one jin xian chariot, one yellow battle ax chariot, one white heron chariot, one leopard tail chariot, one yao chariot, one ram chariot each, five animal skin chariots, and twelve shu chariots. 『大定十一年，將有事於南郊，命太常寺檢宋南郊禮，鹵簿當用玉輅、金輅、象輅、革輅、木輅、耕根車、明遠車、指南車、記里鼓車、崇德車、皮軒車、進賢車、黃鉞車、白鷺車、鸞旗車、豹尾車、軺車、羊車各一，革車五，屬車十二。』

40. San Cai Tu Hui《三才圖會》[25]

[D]ecorations on the right side of the chariot. Using the shu measuring instrument, the height was one chi, four cun and two fen, and the length at the bottom was seven cun and four fen. The linchpin had a diameter of three cun and seven fen. Diameter of the vertical wooden rod was three cun and four fen. A human figure was carved out of jade and its hand pointed to the south. At the foot of the figure was a circular hole for rotating axle; it was placed on top of the picture of Chi You. During the middle Yan You period, it was displayed in the Yao Mu nunnery. 『… 右車飾。以黍尺度高一尺四寸二分，下長七寸四分。轄木口圓徑三寸七分，管立木口圓徑三寸四分。琢玉為人形，手常指南。足底通圓竅，作旋轉軸，踏於蚩尤之上。延祐中獲觀於姚牧庵承旨處。』

7.2 Historical Development

According to legend and historic records [4–9], it was said that both the Yellow Emperor (黃帝, ~2697–2599 BC) and Zhou Gong (周公, ~1122–1035 BC) successfully invented south-pointing chariots. However, they were not recorded in official literature and there was not enough evidence to support the argument. In the Han Dynasty (206 BC–AD 220), several references [9–11] supported the fact that south-pointing chariots were designed. South-pointing chariots appeared in some official literature from the time of the Three Kingdoms (AD 220–280) to the Jin Dynasty (AD 1115–1234). During these periods, over 20 books and other official literature mentioned that south-pointing chariots were successfully designed and manufactured. Moreover, a solid south-pointing chariot designed by Ma Jun (馬鈞) first appeared in the era of Three Kingdoms [9, 12, 13]. There were two detailed records about the exterior shape and the interior structure of south-pointing chariots in the publication Song Shi《宋史》, that is, History of the Song Dynasty, including one design by Yan Su (燕肅) in AD 1027 and another by Wu De-ren (吳德仁) in AD 1107 [18]. Initially, a south-pointing chariot might have been used for military purposes, but in later periods, it became a chariot in an imperial convoy to show the power and prestige of the emperor. Also, the size of south-pointing chariots gradually became larger. Furthermore, no records regarding south-pointing chariots were found after the Yuan Dynasty (AD 1206–1368).

Figure 7.1 shows a device named the south-pointing chariot in the publication San Cai Tu Hui《三才圖會》compiled by Wang Qi (王圻) in AD 1607 [25]. Though a chariot, this device had no wheels. Since the applications of magnetic needles (compass) were popular in the Ming Dynasty (AD 1368–1644), this design might have been used as a toy with a hidden magnet inside to indicate the south–north direction.

The development of south-pointing chariots was not a series of improvement from an existing design. The designs that appeared in different dynasties were likely to have been invented independently. The objects were always destroyed or lost in wars during dynasty changes. So far, no relevant ancient objects or archaeological relics have been found, and no publication could clearly provide any inner ancient mechanism. As a result, the structure of mechanisms of south-pointing chariots in ancient China cannot be assured and their design remains a mystery.

In the 18th century, scholars began to study the existence of south-pointing chariots in ancient China. In early days, people often confused the chariot with the compass, and they even believed that the south-pointing chariot operated via a magnet hidden inside. In 1732, A. Gaubil [26] and other scholars in Europe assumed that the south-pointing chariot was equivalent to the compass. In 1834, J. Klaproth [27] misunderstood and translated the "south-pointing chariot" into "char magnetique," thinking that its wooden man whose finger pointed to the south was by means of a magnet concealed in its body. In 1908, F. Hirth [28] doubted the feasibility of controlling the output to fix the same direction by the use of several gears. He wrote: "... It appears that all that was turned out was a machine consisting of certain wheels, possibly registering the movements of the axle of a chariot in such a manner as to cause an index to point in the same direction, whatever direction the chariot might take. I do not know whether such a construction is actually within the range of possibility."

Figure 7.1 South-pointing chariot in *San Cai Tu Hui* 《三才圖會》 [25]

In 1909, H. A. Giles [29] translated into English two paragraphs of descriptions of Yan Su's south-pointing chariot in the publication Song Shi, that is, History of the Song Dynasty, but failed in his attempts to manufacture the model. In 1925, A. C. Moule [30] retranslated the descriptions into a more exact version and successfully reconstructed Yan Su's south-pointing chariot invented in the Song Dynasty (AD 960–1279). Figure 7.2 shows such a concept. It was a fixed-axis design with an automatic clutch. As the chariot moved and turned, the shaft pulled a rope with a fixed pulley to control the small gear that was adjacent to the central large gear and the vertical gear. The later works of M. Hashimoto [31] in 1926 and Y. Mikami [32] in 1928 further supported Moule's concept.

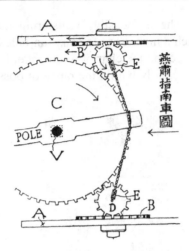

Figure 7.2 Moule's concept of Yan Su's south-pointing chariot [30]

In 1937, Wang Zhen-duo (王振鐸) [33] organized and analyzed various historical writings in ancient China, improved Moule's concept, and built a physical model of Yan Su's south-pointing chariot as shown in Figure 7.3. In Moule's concept, the pole was above the large central horizontal wheel, while in Wang's model the pole was below the large wheel with a supporting shelf between the two parts to balance the large wheel and to facilitate the movement of the pole.

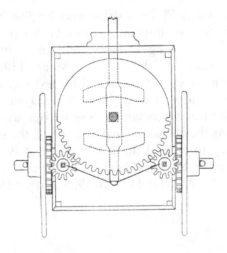

Figure 7.3 Wang's design of Yan Su's south-pointing chariot [33]

Around 1924, K. T. Dykes proposed the idea of using a differential gearing system for south-pointing chariots. Dykes pointed out that Moule's argument was slow and complicated and only a differential gearing system could arrive at the advantages of easy controlled and high accuracy. However, Dykes also admitted that there is no evidence to prove the theory of a differential gearing system.

In 1947, G. Lanchester [34] built a physical model of a south-pointing chariot with a differential gearing mechanism, Figure 7.4. Lanchester bypassed historical materials and suggested that the interior structure of south-pointing chariots should be of similar mechanism to the differential gearing transmissions in automobiles, to permit one wheel to run faster than the other when turning. Perhaps such a hypothesis is another solution to the mystery of south-pointing chariots. However, the concept of a differential gearing system was not found in applications in ancient China. Thereafter, J. Needham [35], S. H. Li [36], and J. Y. Lu [37] summarized studies in south-pointing chariots while some scholars focused on designing different interior mechanisms [38–46, 48–51].

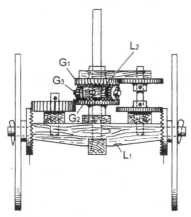

Figure 7.4 Lanchester's design of the south-pointing chariot [34]

7.3 Recent Development

There were two directions in the study of south-pointing chariots taken in the past years. One emphasizes textual criticism of records of ancient literature and recovery [30, 31, 33, 48, 49]; the other hypothesizes that south-pointing chariots existed and accordingly designed the inner mechanisms based on their characteristics and functions [38–46, 50, 51]. Regarding the former, only Wang Zhen-duo had successfully recovered Yan Su's south-pointing chariot based on the statements in Song Shi, that is, History of the Song Dynasty. For the latter, many scholars have designed individual south-pointing chariots with mechanisms that included gears, linkages, ropes, pulleys, and friction wheels.

In 1954, Liu Xian-zhou [47] criticized the designs of Moule and Wang because the description of Yan Su's south-pointing chariot did not mention the ropes and pulleys. In 1962, he [48] pointed out that Bao Si-he's (鮑思賀) design of Yan Su's south-pointing chariot in 1948, Figure 7.5, was more reasonable. In 1977, A. C. Sleeswyk [49] proposed a fixed-axis design of Yan Su's south-pointing chariot with an automatic clutch and gears, ratchets, pallets, and pole, Figure 7.6. While turning around, the ratchet and the detent mesh the gears through the pole arrangement and transmit motion to the output. This design was extremely complex and the components were not found in applications in ancient China.

In recent years, some scholars have provided a series of designs especially based on differential gear trains with no consideration to historical literature. Such designs are accurate in determining direction, simple in operation, and concise in mechanism.

In 1979, Z. M. Lu proposed three designs of south-pointing chariots [38], Figure 7.7, one of which is similar to Lanchester's concept. In 1982, Z. R. Yan emphasized that in the book Song Shu 《宋書》, that is, Book of the Song Dynasty, regarding Zu Chong-zhi's (祖沖之) south-pointing chariot, it can only be achieved with a differential gearing system: "... the chariot was refined, and no adjustments were necessary even after hundreds of rotation." 『 ... 其制甚精，百曲千回，未常移變。 』. He also provided two designs as shown in Figure 7.8 [39]. In 1986, Y. Z. Yang designed two south-pointing chariots [41], Figure 7.9. In 1990, M. Muneharu and K. Satoshi showed a design based on differential gear trains with 16 spur gears, Figure 7.10 [42, 43]. In 1996, L. C. Hsieh, J. Y. Jen, and M. H. Hsu synthesized the design of south-pointing chariots for arbitrary planetary gear trains with two degrees of freedom [45], Figure 7.11. In 1999, Y. J. Chen provided a design with rollers and ropes instead of spur and bevel gears as in Lanchester's model, Figure 7.12 [46]. In 2006, H. S. Yan and C. W. Chen synthesized all possible design configurations of south-pointing chariots with differential design specifications and requirements [1], and Figure 7.13 shows one of the feasible designs.

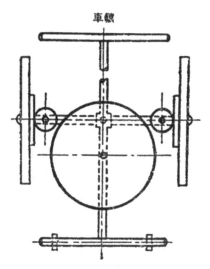

Figure 7.5 Bao's design of Yan Su's south-pointing chariot [48]

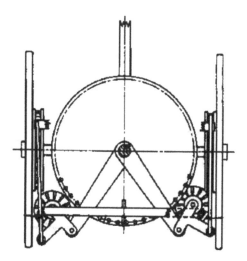

Figure 7.6 Sleeswyk's design of Yan Su's south-pointing chariot [49]

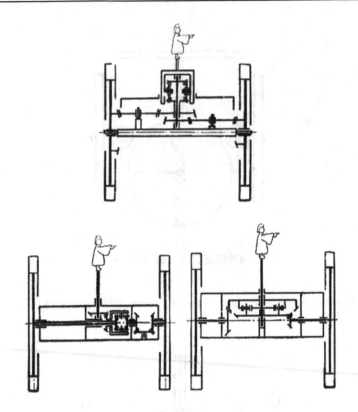

Figure 7.7 Lu's south-pointing chariots [38]

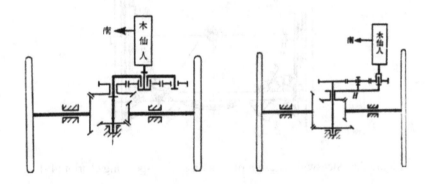

Figure 7.8 Yan's south-pointing chariots [39, 40]

Figure 7.9 Yang's south-pointing chariots [41]

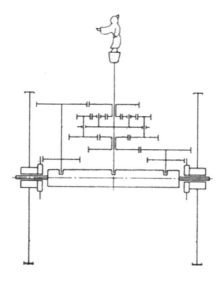

Figure 7.10 Muneharu and Satoshi's south-pointing chariot [42, 43]

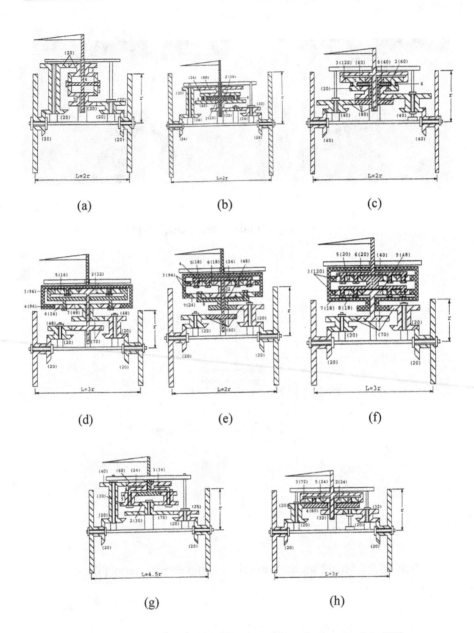

Figure 7.11 Hsieh, Jen, and Hsu's south-pointing chariots [45]

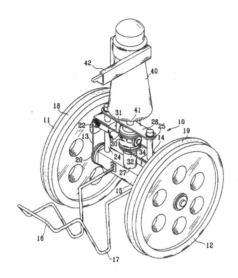

Figure 7.12 Chen's south-pointing chariots [46]

Figure 7.13 Yan and Chen's south-pointing chariot [1]

7.4 Topological Structures

In 1994, J. Y. Lu [37] classified south-pointing chariots into two types: fixed-axis type and differential type according to historical records and inner mechanisms. The fixed-axis-type south-pointing chariots are close to historical descriptions in Song Shu, that is, Book of the Song Dynasty, with one degree of freedom. But this type of chariot is difficult to control. Instead, differential-type south-pointing chariots provide better performance and accuracy, and the degrees of freedom are two. However, such a design or relative applications was not discovered in ancient China.

Functionally, existing designs of south-pointing chariots can be divided into four parts including two inputs, a transmission mechanism, a passive feedback mechanism, and an output, Figure 7.14 [1].

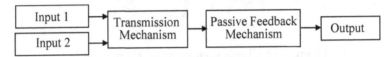

Figure 7.14 Decomposition of south-pointing chariots

Inputs
According to the historical records of ancient China, south-pointing chariots had two wheels as the inputs. When the chariot makes a straight-line motion, the two inputs have the same angular velocity and the body fixes the direction. When the chariot turns left or right, the angular velocities of the two inputs are different, causing the body to rotate.
Transmission mechanism
The transmission mechanism connects the two inputs and the passive feedback mechanism. There are two types: one directly connects the input members with the wheels, such as the design of Z. R. Yan [39]. The other adds some components for transmitting motions, such as the design of M. Muneharu and K. Satoshi [43].
Passive feedback mechanism
The passive feedback mechanism is the main part of a south-pointing chariot. In the fields of cybernetics and natural science, there are two types of feedback: positive feedback and passive (negative) feedback. The former enlarges the input signals, and the latter balances the whole system. The passive feedback mechanism receives the two inputs, and outputs the same rotational angle of the carriage in the opposite direction. From this perspective, J. Needham [35] suggested that the south-pointing chariot was the first cybernetic device in the world.

Output

The output is a member of the passive feedback mechanism that shows the fixed direction. According to historical records of ancient China, there was a wooden figure of a person on the output link whose finger pointed to the desired fixed direction.

The process for analyzing the topological structures of south-pointing chariots is as follows, Figure 7.15 [1].

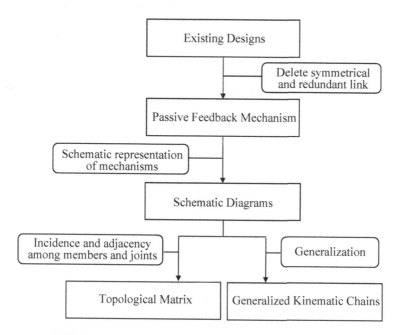

Figure 7.15 Analysis of topological structures of south-pointing chariots

Step 1. Analyze existing designs to differentiate the inputs, the transmission mechanism, the passive feedback mechanism, and the output.

Step 2. Delete redundant links and symmetrical links from the passive feedback mechanism obtained in Step 1 to simplify the number of links and joints.

Step 3. Draw the corresponding schematic diagram from the result obtained in Step 2 and assign numbers to all links and joints.

Step 4. Transform the schematic diagram obtained in Step 3 into its corresponding generalized kinematic chain.

Step 5. List the topology matrix of the schematic diagram obtained in Step 3 or the generalized kinematic chain obtained in Step 4 to recognize the incidence and adjacency among members and joints.

Again, Figure 7.16(a) shows the south-pointing chariot designed by George Lanchester, as previously shown in Figure 7.4 [34]. The analysis of this design is as follows:

1. The two inputs are the left wheel K_{wa} and right wheel K_{wa1}. The transmission mechanism includes members K_{Lb}, K_{Lb1}, and K_{Lb2}. The passive feedback mechanism includes members K_{Lc}, K_{Lc1}, K_{Ld}, K_{Ld1}, and K_{Le}. And, member K_{Lf} is the output link.

2. Based on the result in step 1, delete the redundant member K_{Ld1}.

3. Based on the result in step 2, draw the corresponding schematic diagram and then assign numbers to all links and joints, as shown in Figure 7.16(b).

4. Based on the result in step 3, transform the schematic diagram into the corresponding (5, 6) generalized kinematic chain as shown in Figure 7.16(c).

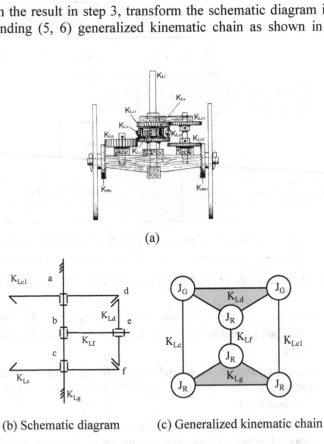

(a)

(b) Schematic diagram   (c) Generalized kinematic chain

Figure 7.16   Lanchester's south-pointing chariot and its generalized kinematic chain [34]

Through the process of analysis, the topological characteristics of all available south-pointing chariots can be concluded. The mechanical components of existing designs include linkages [31, 33, 34, 38–46, 48–51], gears [30, 33, 34, 38–46, 48–51], ropes and pulleys [30, 33, 45], and frictional wheels [45].

Based on the analysis of 25 existing designs, Figure 7.17 shows two topological structures of south-pointing chariots with four members, Figure 7.18 shows seventeen topological structure of south-pointing chariots with five members, Figure 7.19 shows three topological structures of south-pointing chariots with six members, Figure 7.20 shows two topological structures of south-pointing chariots with seven members, and Figure 7.21 shows one topological structure of south-pointing chariot with eight members [45].

Many different designs have the same topological structures after obtaining the corresponding generalized kinematic chains. For the existing south-pointing chariots shown in Figures 7.17–7.21, the designs in the following groups have the same generalized kinematic chains:

1. The designs shown in Figure 7.18 (5), Figure 7.18 (13), and Figure 7.18 (14).
2. The designs shown in Figure 7.18 (11) and Figure 7.18 (12).
3. The designs shown in Figure 7.18 (2), Figure 7.18 (12), and Figure 7.18 (15).

Existing designs of south-pointing chariots with a fixed-axis wheel system are fewer than those with a differential gearing system because the latter are easier to control and the inner constructions are more flexible than the former. From literature studies, only two designs belong to the system with a fixed-axis wheel [33, 48]. As a result, based on the analysis of topological structures, the characteristics of south-pointing chariots with a fixed-axis wheel system are concluded as follow:

1. The topological structures are different between the phases of the straight motion and the turning motion.
2. The topological structures are symmetrical to the line from the output to the frame.
3. The degree of freedom is 1. And, there is a link or a rope that functioned as a clutch when the chariot changes direction.
4. The function of the pulleys is to change the direction of the ropes.

Figure 7.22(a) shows the design by Z. D. Wang [33] and its corresponding topological structure in the form of a generalized kinematic chain. In the phase of turning motion, a shaft pulls the rope to control the up or down movement of the two gears that connect to the output gear and two wheels. Since only half of the mechanism is affected in the turning motion, the generalized kinematic chain can be divided into two identifiable parts along the symmetrical axis as shown in Figure 7.22(b). However, for the sake of simplicity, the pulleys are ignored [1].

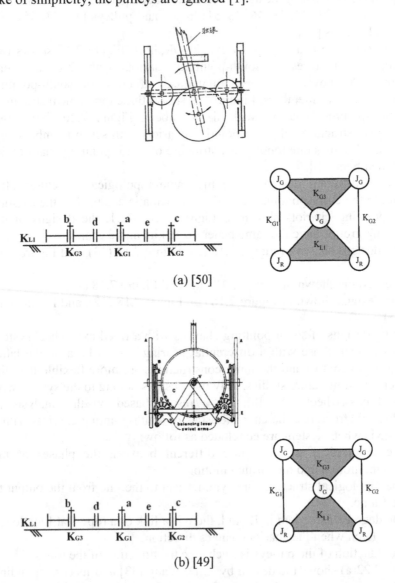

(a) [50]

(b) [49]

Figure 7.17 Topological structures of south-pointing chariots with four members

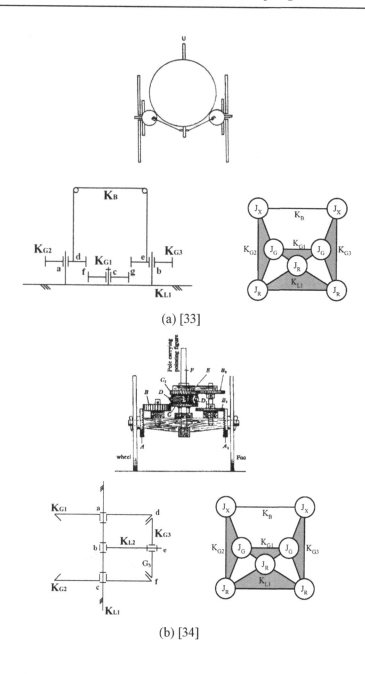

(a) [33]

(b) [34]

Figure 7.18 Topological structures of south-pointing chariots with five members

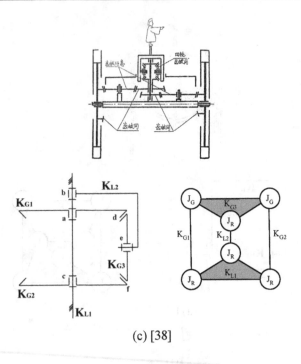

(c) [38]

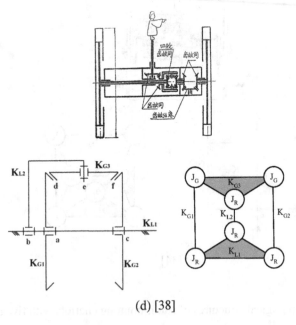

(d) [38]

Figure 7.18 (*Continued*)

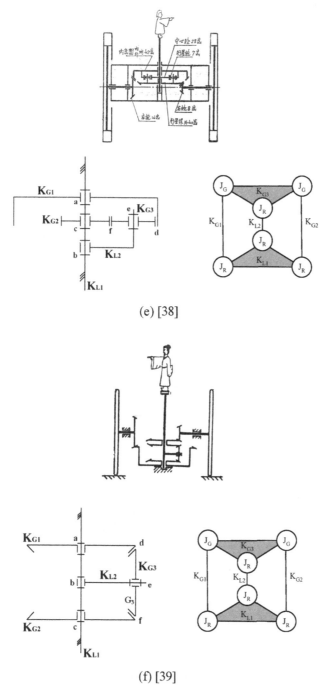

(e) [38]

(f) [39]

Figure 7.18 (*Continued*)

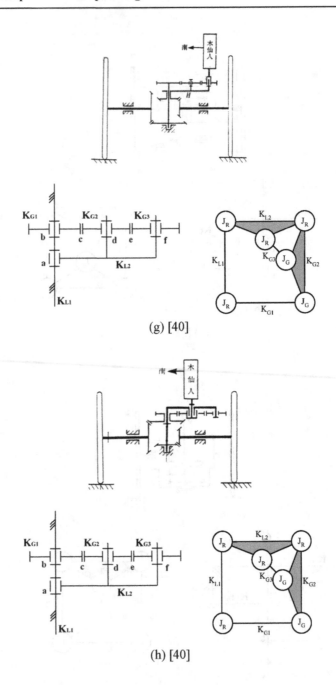

(g) [40]

(h) [40]

Figure 7.18 (*Continued*)

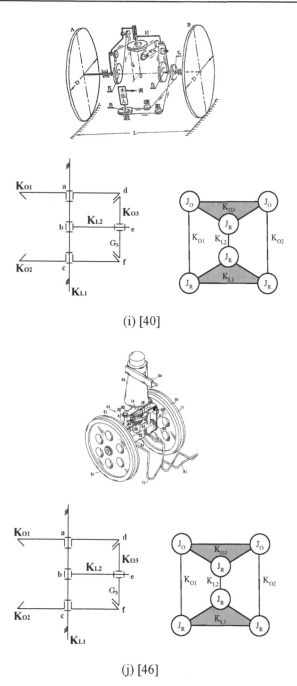

(i) [40]

(j) [46]

Figure 7.18 (*Continued*)

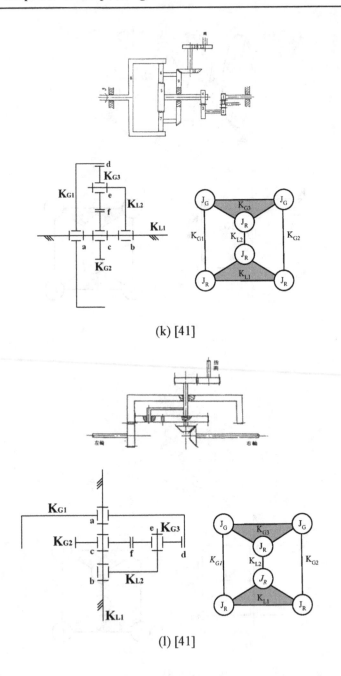

(k) [41]

(l) [41]

Figure 7.18 (*Continued*)

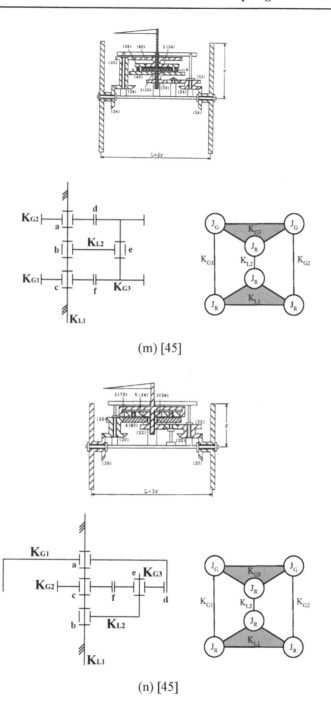

(m) [45]

(n) [45]

Figure 7.18 (*Continued*)

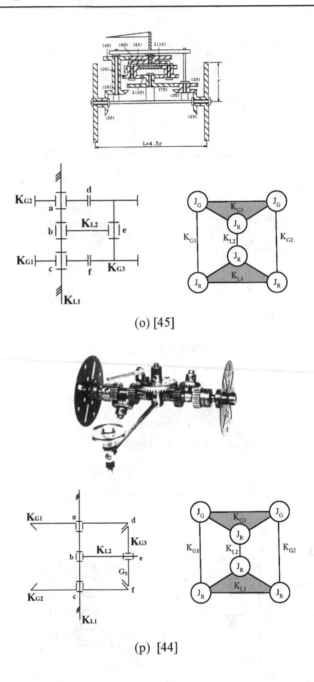

(o) [45]

(p) [44]

Figure 7.18 (*Continued*)

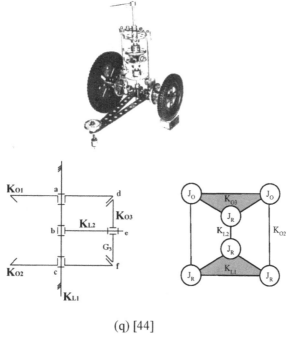

(q) [44]

Figure 7.18 (continued)

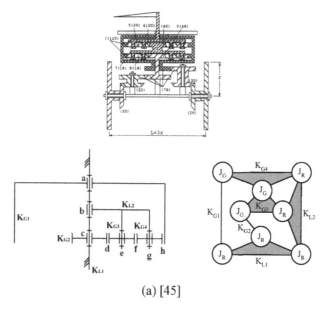

(a) [45]

Figure 7.19 Topological structures of south-pointing chariots with six members

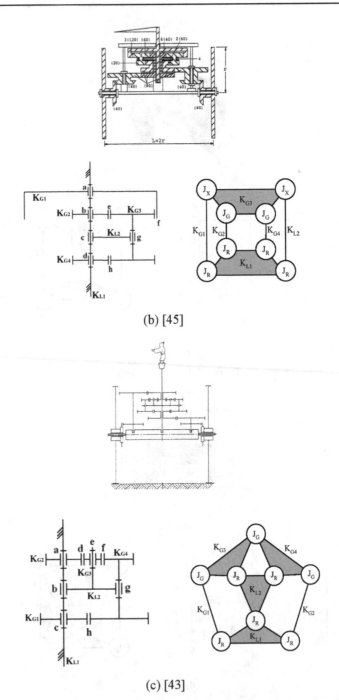

(b) [45]

(c) [43]

Figure 7.19 (*Continued*)

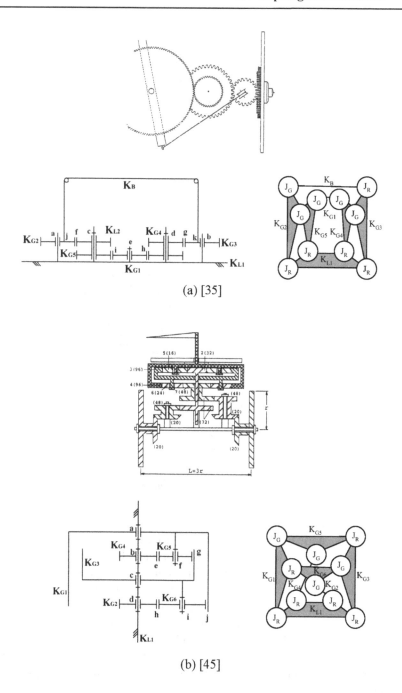

Figure 7.20 Topological structures of south-pointing chariots with seven members

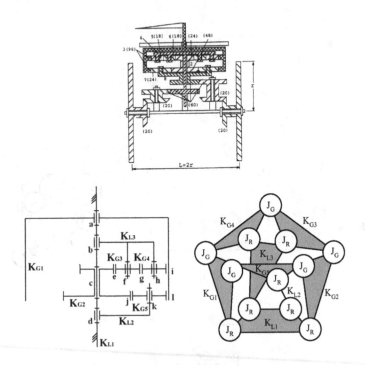

Figure 7.21 Topological structures of south-pointing chariots with eight members [46]

7.5 Representations of Joints and Members

Some of the existing designs have the same generalized kinematic chain, but the inputs and the output are identified as different members. Furthermore, different designs are also obtained by replacing different types of gears and by designing different transmission subsystems. Therefore, detailed representations are needed to identify the joints and members in the generalized kinematic chain of south-pointing chariots.

From the results of analysis, all axial directions of revolute joints in existing mechanisms of south-pointing chariots are either horizontal or vertical to the ground. For nonrevolute joints, there are three connecting types: perpendicular, internal, and external. The representation of joints is as follows:

$$J_{\text{type of joint}}^{\text{characteristic of joint}}$$

in which the subscript denotes the type of joint and the superscript denotes the characteristic of the joint. For revolute joints, there are two characteristics: vertical (D_V) or horizontal (D_H) to the ground. For nonrevolute joints, there are three types: the joint is incident to two members with perpendicular axial direction (G_p), the joint is incident to two members with parallel axial direction and with external connection (G_e) or internal connection (G_i), respectively. For examples, the joint denoted as $J_R^{D_v}$ indicates that the axial direction of a revolute joint is vertical to the ground, and the joint denoted as $J_G^{G_p}$ indicates that the characteristic of a gear joint is incident to two gears whose axial directions are perpendicular to each other. However, it is not necessary to identify any particular characteristic for a fixed joint, i.e, a joint without any relative motion.

Although the passive feedback mechanism is the key in a south-pointing chariot, the design of the inputs, the output, and the transmission mechanism are also indicated in the final results. In fact, some existing designs have the same passive feedback mechanism but with different inputs, transmission parts, and output. This information should be included in the topological structure. The representation of members is as follows:

$$K_{\text{type of member}}^{\text{input/output, type of transmission part}}$$

in which the subscript denotes the type of members and the superscript denotes the input/output and type of transmission part. For examples, a joint denoted as $K_G^{I1,G}$ indicates a gear as input 1 and that the transmission part is composed of gears, while a joint denoted as $K_O^{I2,D}$ indicates a roller as input 2 and that it is directly adjacent to the wheel.

Figure 7.22(c) shows the corresponding generalized kinematic chain of Wang's design based on the above-mentioned representations of joints and members. Figure 7.23(a) and (b) show the corresponding generalized kinematic chains of Lanchester's design and Hsieh's design, respectively. The advantage of such representations of members and joints is that the designer can transform all feasible designs into different generalized kinematic chains in the process of analysis, while retaining information on the inputs, the output, and the transmission mechanism. In addition, such representations are also useful for systematically synthesizing all possible design concepts of south-pointing chariots.

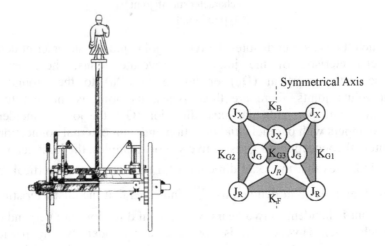

(a) Wang's design and its topological structure

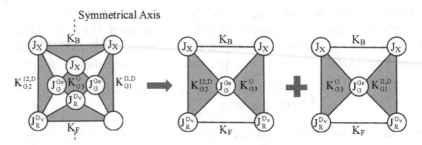

(b) Dividing into two identical parts along the symmetrical axis

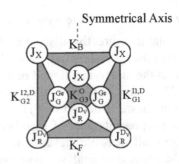

(c) Topological structure with particular identities

Figure 7.22 Analysis of Wang's south-pointing chariot

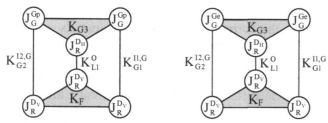

Figure 7.23 Generalized kinematic chains of Lanchester's and Hsieh's designs with new representations

7.6 Reconstruction Design

Figure 7.24 shows the design procedure for synthesizing the topological structure of south-pointing chariots. It consists of the following five steps:

Step 1. Design specifications

The mechanisms of south-pointing chariots might have various components in different dynasties based on the ancient scientific theories and technologies of that time. Therefore, the first step of the design process is to define the design specifications, including the types and numbers of members and joints, and to conclude the topological characteristics.

Step 2. Generalized kinematic chains

According to the design specifications defined in Step 1, the atlas of generalized kinematic chains with the required numbers of links and joints can be generated based on the algorithm of number synthesis [52], or simply identified from Section 4.4.

Step 3. Specialized chains

Through the process of specialization, specific types of members and joints are assigned to every generalized kinematic chain available in Step 2 to obtain the corresponding atlas of specialized chains subject to design requirements and constraints concluded from the topological characteristics.

Step 4. Specialized chains with particular identities

The axial direction of each revolute joint and connection characteristic of each nonrevolute joint are assigned first. The members of the inputs and the output are identified, and a suitable type of the transmission mechanism with the atlas of specialized chains derived in Step 3 is combined to generate the atlas of specialized chains with particular identities.

Step 5. Reconstruction designs

According to the requirements of motion and function of the designs, the corresponding schematic formats are particularized from the atlas of specialized chains with particular identities to establish the atlas of reconstruction designs.

In what follows, various examples for the reconstruction design of south-pointing chariots in different dynasties of ancient China are presented according to the process shown in Figure 7.24.

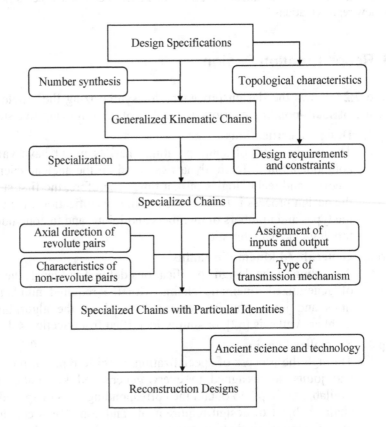

Figure 7.24 Process of reconstruction design

7.7   Differential types with Four Members

In the process of mechanism design, engineers always accomplish the motion and functional requirements with fewer links. Since the south-pointing chariot is composed of two inputs, one output and a frame, the number of links is at least four. Therefore, the differential-type south-pointing chariots with four links are synthesized as follows (Example 7.1).

Step 1. Design specifications
The type of member is open in this case. The design specifications are:
1. The number of links of the passive feedback mechanism is four.
2. The degree of freedom is two.
3. Mechanical components are linkages, gears, ropes, pulleys, and rollers.

Step 2. Generalized kinematic chains
For a planar mechanism with two degrees of freedom ($F_p = 2$) and four links ($N_L = 4$), the number of joints is 4 ($N_J = 4$; one joint with two degrees of freedom and three joints with one degree of freedom) or 5 ($N_J = 5$; three joints with two degrees of freedom and two joints with one degree of freedom). Figure 7.25(a) and (b) show the atlas of generalized kinematic chains with four links and four and five joints, respectively.

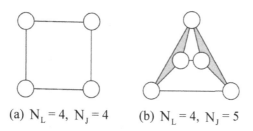

(a) $N_L = 4$, $N_J = 4$ (b) $N_L = 4$, $N_J = 5$

Figure 7.25 Atlas of generalized chains (Example 7.1)

Step 3. Specialized chains
Once the atlas of generalized kinematic chains is obtained, all possible specialized chains can be identified according to the following substeps:
1. For each generalized kinematic chain, identify the frame link for all possible cases.
2. For each case obtained in substep 1, assign revolute joints.
3. For each case obtained in substep 2, assign nonrevolute joints.
The members and the joints must be assigned subject to the following design requirements and constraints:
Frame (K_F)
1. One of the links in each generalized kinematic chain must be the frame.
2. A frame must not be included in a three-bar loop in the chain.
3. A frame must be a multiple link in order to have two input members and one output member.

Revolute joint (J_R)
1. There must be $N_L - 1$ revolute joints.
2. Any joint incident to the frame must be a revolute joint.
3. Every link must have at least one revolute joint.
4. There can be no loop formed exclusively by revolute joints.
Rolling joint (J_O) *or gear joint* (J_G)
1. A binary link cannot have two gear or rolling joints.
2. A ternary link can only have two gear or rolling joints.
3. There can be no three-bar loop formed exclusively by gear joints.
After assigning all types of joints and members, the atlas of specialized chains can be obtained as shown in Figure 7.26.

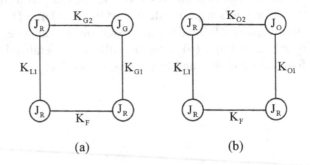

(a)   (b)

Figure 7.26 Atlas of specialized chains (Example 7.1)

Step 4. Specialized chains with particular identities
In this step, the axial directions of revolute joints and the characteristics of nonrevolute joints are identified as follows:
1. Identify the superscript "D_V" or "D_H" to each revolute joint.
2. There must be at least one revolute joint in the horizontal direction.
3. The subscript of nonrevolute joints is identified as "G_p" to represent the axial directions of the two adjacent members that are perpendicular to each other.
4. The subscript of nonrevolute joints is identified as "G_i" or "G_e" to represent the axial directions of the two adjacent members that are horizontal. Here, subscript "i" or "e" indicates that the two gears (rollers) incident to a gear (rolling) joint can be internal or external.
Different designs are obtained by assigning different members as the two inputs and the output in the specialized chains as follows:
Input
(a) The input link must be adjacent to the frame.
(b) Input 1 must not be adjacent to input 2.

Output
(a) The axial direction of the output link must be vertical to the ground.
(b) There can be no loop formed exclusively by the two inputs and the output.

Finally, the type of transmission mechanism in all specialized chains is considered. The transmission mechanism connects the two inputs and the passive feedback mechanism. In general, the simplest solution is direct connection. Since the gear train system of south-pointing chariots may add a compound gear as the transmission mechanism, south-pointing chariots with frictional components have two choices: one is to add a compound roller and the other is to connect with pulleys and ropes. As a result, the atlas of specialized chains with particular identities is obtained as shown in Figure 7.27.

Step 5. Reconstruction designs
It should be noticed that the rotational direction of the output should be opposite to the chariot when turning left or right. Figure 7.28 shows the corresponding eight design concepts of differential-type south-pointing chariots with four links for the atlas of specialized chains with particular identities shown in Figure 7.27. And, the one shown in Figure 7.28(d) is the design by Z. R. Yan [39].

7.8   Gear Train Differential types with Five Members

Available literature indicates that most of the feasible mechanisms of the differential-type south-pointing chariots are gear train systems with five members. According to archeological study, it seems that gears were applied to some mechanical devices during the Qin Dynasty (221–206 BC) and the Han Dynasty (206 BC–AD 220). Furthermore, many agricultural machines in ancient China used gears. Hence, most scholars agree that the south-pointing chariots were designed with gear train systems in ancient China. Therefore, the goal of this design example (Example 7.02) is to generate all possible designs of the differential-type south-pointing chariots with five members and with gear train systems. Thus, the design specifications are as follow:
1. The number of links of the passive feedback mechanism is five.
2. The degree of freedom is two.
3. Mechanical components are links and gears.

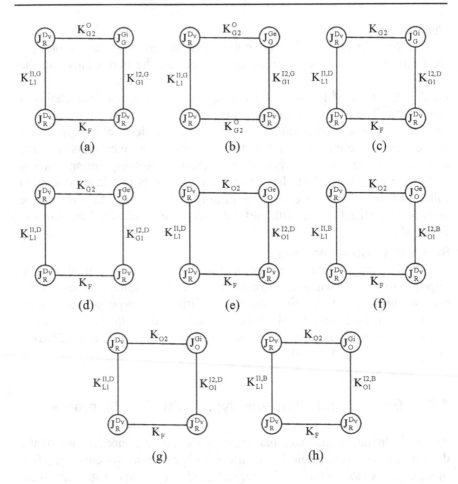

Figure 7.27 Atlas of specialized chains with particular identities (Example 7.1)

Following the same design procedure shown in Figure 7.24 and described in Example 7.1, the corresponding atlas of generalized kinematic chains can be obtained according to the theory of number synthesis. There are two (5, 6) generalized kinematic chains and one (5, 7) generalized kinematic chain as shown in Figure 7.29. Besides, the numbers of gear joints and revolute joints are calculated in this step. The next step is to generate feasible specialized chains. Since Figure 7.29(a) and (c) do not satisfy the design requirements and constraints, there is only one feasible specialized chain as shown in Figure 7.30 which contains two gear joints and four revolute joints. Next, the axial directions of revolute joints and characteristics of gears need to be identified to generate the atlas of specialized chains with particular identities. For example, Figure 7.31(b_1)–(b_4)

show that the axial directions of the four revolute joints are vertical to the ground, and the characteristics of the two gear joints are external conjugation. Then, there are two possible assignments to the two inputs and the one output. One is that the two gears are assigned as the two inputs and the link is assigned as the output; the other one is that one gear and the link are assigned as the two inputs and another gear is assigned as the output. Regarding the transmission mechanisms, either a direct or compound connection of gears can be utilized. Therefore, the atlas of specialized chains with particular identities is obtained as shown in Figure 7.31. According to the procedure of particularization, Figure 7.32 shows the atlas of 18 reconstruction designs in which Figure 7.32(d_1) is Lanchester's design [34], Figures 7.32(a_4) and 7.32(d_4) are Lu's designs [38], Figure 7.32(d_3) is Yan's design [40], Figure 7.32(a_5) is Yang's design [41], and Figure 7.32(a_3), (b_1), and (b_2) are Hsieh's designs [45].

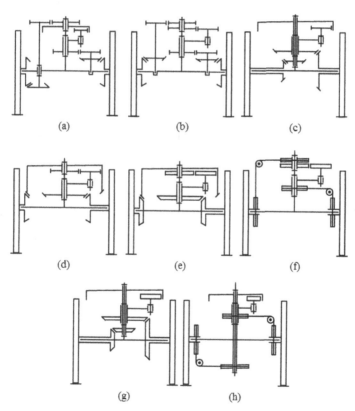

Figure 7.28   Atlas of reconstruction designs (Example 7.1)

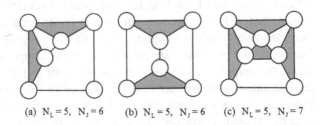

(a) $N_L = 5$, $N_J = 6$ (b) $N_L = 5$, $N_J = 6$ (c) $N_L = 5$, $N_J = 7$

Figure 7.29 Atlas of generalized kinematic chains (Example 7.2)

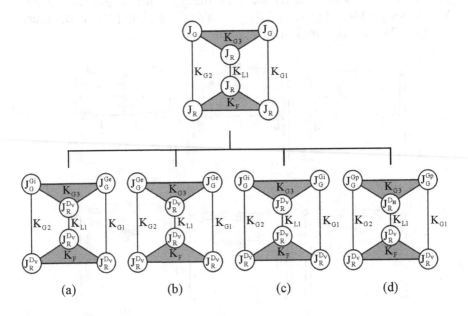

(a)   (b)   (c)   (d)

Figure 7.30 Atlas of specialized chains (Example 7.2)

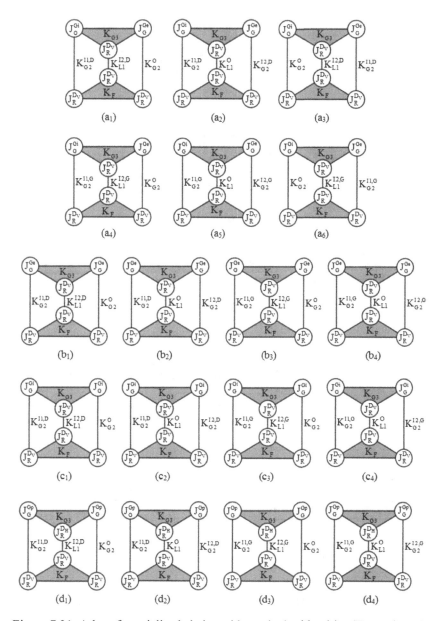

Figure 7.31 Atlas of specialized chains with particular identities (Example 7.2)

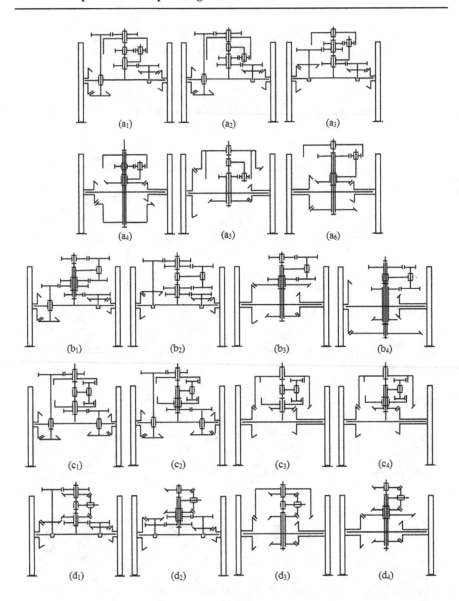

Figure 7.32 Atlas of reconstruction designs (Example 7.2)

7.9 Differential types with Ropes, Pulleys, and Friction Wheels

In ancient China, the development of labor-saving devices matured and found various applications, as presented in Section 3.2, especially the rope-and-pulley mechanisms. Winches were very popular and widely used before the Qin Dynasty (221–206 BC). Moreover, friction wheels have the function of transmitting continuous rotational motion and the advantage of being simple in structure. In this example (Example 7.3), differential-type south-pointing chariots with ropes, pulleys, and friction wheels are synthesized. Thus, the design specifications are as follow:
1. The number of links of the passive feedback mechanism is five.
2. The degrees of freedom are two.
3. The mechanical components are ropes, pulleys, and friction wheels.
For practical applications, the two friction wheels adjacent to a rolling joint have only an external connection. The rope-and-pulley is applied as the transmission mechanism.

Following the same design procedure as shown in Figure 7.24 and described in Example 7.1, the corresponding atlas of generalized kinematic chains can be obtained according to the theory of number synthesis. There are two (5, 6) generalized kinematic chains and one (5, 7) generalized kinematic chain shown in Figure 7.33. Furthermore, the numbers of rolling pairs and revolute pairs are calculated in this step. The next step is to generate feasible specialized chains. Since Figures 7.33(a) and (c) cannot satisfy the design requirements and constraints, the number of the atlas of specialized chains is only one, Figure 7.34. Next, the axial directions of revolute pairs and characteristics of gears need to be identified to generate the atlas of specialized chains with particular identities. For example, Figure 7.34(a) shows that the axial directions of the four revolute pairs are vertical to the ground, and the characteristics of the two rolling pairs have external connection. Then, there are two possible assignments to the two inputs and the one output. One is that the two rollers are assigned as the two inputs and the link is assigned as the output, the other one is that one roller and the link are assigned as the two inputs and another roller is assigned as the output. Regarding the transmission mechanisms, either a direct connection or with ropes and pulleys can be utilized. Hence, the atlas of specialized chains with particular identities as shown in Figure 7.35 is obtained. According to the procedure of particularization, Figure 7.36 shows the atlas of four reconstruction designs in which Figure 7.36(b) is Y. J. Chen's concept [46].

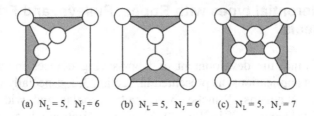

(a) $N_L = 5$, $N_J = 6$ (b) $N_L = 5$, $N_J = 6$ (c) $N_L = 5$, $N_J = 7$

Figure 7.33 Atlas of generalized kinematic chains (Example 7.3)

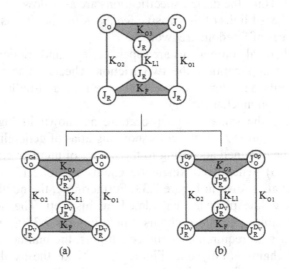

Figure 7.34 Atlas of specialized chains (Example 7.3)

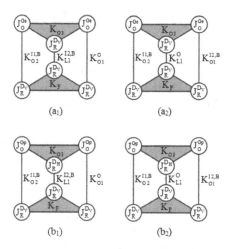

Figure 7.35 Atlas of specialized chains with particular identities (Example 7.3)

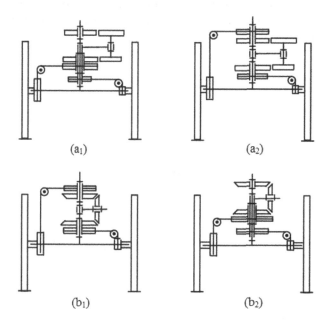

Figure 7.36 Atlas of reconstruction designs (Example 7.3)

7.10   Fixed-axis types with Three Members

Since the half configuration of a south-pointing chariot is composed of one input, one output, and the frame, the number of links is three at least. This example (Example 7.4) synthesizes the fixed-axis wheel system south-pointing chariots with three members as follows.

Step 1. Design specifications
The type of member is open in this case, and the design specifications are:
1. The number of links of the half passive feedback mechanism is three.
2. The degree of freedom is one regardless of going straight or turning a corner.
3. The mechanical components are linkages, gears, and rollers.

Step 2. Generalized kinematic chains
For a planar mechanism with one degree of freedom ($F_p = 1$) and three links ($N_L = 3$), the number of joints is three ($N_J = 3$; one joint with two degrees of freedom, and two joints with one degree of freedom). Figure 7.37(a) shows the atlas of generalized kinematic chains with three links and three joints.

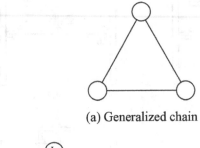

(a) Generalized chain

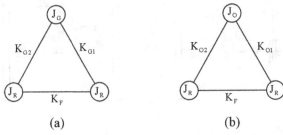

(a)   (b)

Figure 7.37   Atlases of generalized kinematic chains and specialized chains (Example 7.4)

Step 3. Specialized chains
Once the atlas of generalized kinematic chains is obtained, all possible specialized chains can be identified according to the following substeps:
1. For each generalized chain, identify the frame link for all possible cases.
2. For each case obtained in substep 1, assign revolute joints.
3. For each case obtained in substep 2, assign nonrevolute joints.
These steps are carried out subject to the following design requirements and constraints:
Frame (K_F)
1. One of the links in each generalized kinematic chain must be the frame.
Revolute joint (J_R)
1. Any joint incident to the frame must be a revolute joint.
2. Every link must have at least one revolute joint except for belts/ropes.
3. No loop can be formed exclusively by revolute joints
Gear joint (J_G) *or rolling joint* (J_O)
1. A binary link cannot have two gear joints or rolling joints.
2. A ternary link can only have two gear joints or rolling joints.
3. No three-bar loop can be formed exclusively by gear joints.
After assigning all types of joints and members, two specialized chains are obtained as shown in Figure 7.37(b).

Step 4. Specialized chains with particular identities
In this step, the axial directions of revolute joints and the characteristics of nonrevolute joints are identified first according to the following points:
1. Identify the superscript "D_V" or "D_H" of each revolute joint.
2. There must be at least one axial direction of revolute joint as the vertical direction.
3. The subscripts of nonrevolute joints are identified as "G_p" to represent the axial directions of the two adjacent members that are perpendicular to each other.
4. The subscript of nonrevolute joints is identified as "G_i" or "G_e" to represent the axial directions of the two adjacent members that are horizontal. Here, subscript "i" or "e" indicates that the two gears (rollers) adjacent to a gear (rolling) joint can be internal or external.
Different designs can be obtained by assigning different members as the two inputs and the output in the specialized chains. The inputs and the output are identified as follows:
Input
1. The input link must be adjacent to the frame.
Output
1. The axial direction of the output link must be vertical to the ground.
2. The output link must be adjacent to the frame.

Finally, the type of transmission subsystem in all the specialized chains is considered. The transmission subsystem is adjacent to the two inputs and the passive feedback mechanism. Here, the simplest solution, a direct connection, is chosen. As a result, six specialized chains with particular identities are obtained as shown in Figure 7.38.

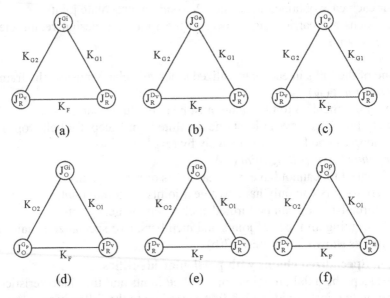

Figure 7.38   Atlas of specialized chains with particular identities – half of the design (Example 7.4)

Step 5. Reconstruction designs
The two identical specialized chains with particular identities obtained in the above step are combined, Figure 7.38, along the symmetrical axis to obtain the complete configurations of the designs shown in Figure 7.39. Also, the rotational direction of the output should be opposite to the chariot when turning left or right.

Figure 7.40 shows the corresponding six design concepts of the fixed-axis wheel system south-pointing chariots with three links for the atlas of specialized chains with particular identities shown in Figure 7.39. The one shown in Figure 7.39(d_1) was designed by S. H. Bao [48].

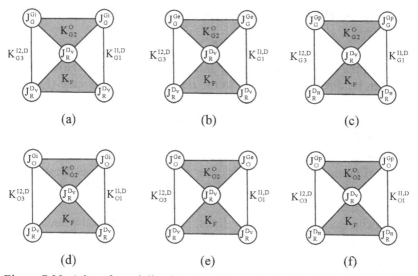

Figure 7.39 Atlas of specialized chains with particular identities (Example 7.4)

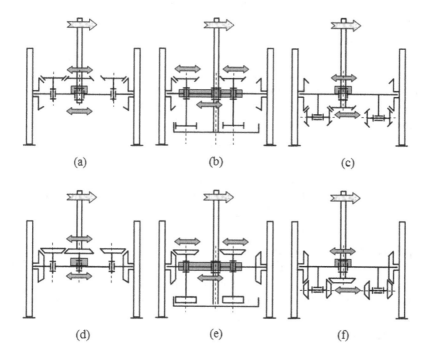

Figure 7.40   Atlas of reconstruction designs (Example 7.4)

7.11 Fixed-axis types with Two Members

According to the analysis of existing designs, there are at least three members in the half configuration of south-pointing chariots. However, the wheels can be directly connected to the output link. In this case (Example 7.5), the design specification that there must be one link as the input can be removed to synthesize the simplest designs, that is, the fixed-axis wheel south-pointing chariots with two members.

Step 1. Design specifications
The type of member is open in this case. And, the design specifications are:
1. The number of links of the passive feedback mechanism is two.
2. The degree of freedom is one.
3. The mechanical components are links, gears and frictional wheels.

Step 2. Generalized kinematic chains
For a planar mechanism with one degree of freedom ($F_p = 1$) and two links ($N_L = 2$), the number of joints is one ($N_J = 1$; one joint with two degrees of freedom). Figure 7.41(a) shows the generalized kinematic chains with two links (open chain).

Step 3. Specialized chains
For generating the specialized chains, the two links are assigned to the frame and the output link respectively due to the reasons that the input links are eliminated and one joint is assigned as the revolute joint. The output link can be a gear or a frictional wheel. Finally, two specialized chains are obtained as shown in Figure 7.41(b).

Step 4. Specialized chains with particular identities
There is only one way to assign the horizontal axis to the revolute joint of the output. The transmission mechanism connects the two inputs and the passive feedback mechanism. Here the simplest solution, a direct connection, is chosen. As a result, there are two specialized chains with particular identities as shown in Figure 7.41(c).

Step 5. Reconstruction designs
In this case, the two identical specialized chains with particular identities along the symmetrical axis do not need to be combined, since the output connects directly to the two wheels. Figure 7.41(d) shows the atlas of two reconstruction designs. These two designs have the least number of links and are the simplest mechanisms.

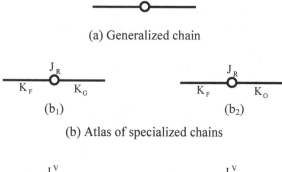

(a) Generalized chain

(b_1) (b_2)

(b) Atlas of specialized chains

(c_1) (c_2)

(c) Atlas of feasible specialized chains with particular identities

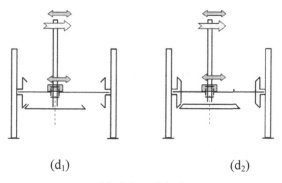

(d_1) (d_2)

(d) Atlas of designs

Figure 7.41 Simplest designs (Example 7.5)

7.12 Fixed-axis types with Various Elements

Yan Su's (燕肅) south-pointing chariot, reconstructed by Z. D. Wang, contains ropes and pulleys for pulling the gears. In fact, in ancient China, the developments of labor-saving devices were very mature and had various applications, especially the rope-and-pulley mechanisms. Moreover, winches were widely used before the Qin Dynasty (221–206 BC). Besides, the friction wheels have the function of transmitting continuous rotational motion and the advantage of simplicity in structure. In this case (Example 7.6), the fixed-axis wheel south-pointing chariots with ropes, pulleys, pulleys, gears, linkages, and friction wheels are synthesized.

Step 1. Design specifications

The type of member is open in this case, and the design specifications are:

1. The number of links of the passive feedback mechanism is four.
2. The degree of freedom is one.
3. The mechanical components are links, gears and frictional wheels.

Step 2. Generalized kinematic chains

The one rope and two fixed joints are ignored since they do not function when the south-pointing chariot moves straight ahead. First, for a planar mechanism with one degree of freedom ($F_p = 1$) and four links ($N_L = 4$, includes three members and one rope), the number of joints is five ($N_J = 5$; one joint with two degrees of freedom, two joints with one degree of freedom, and two fixed joints). Therefore, the generalized kinematic chain is four links and five joints as shown in Figure 7.42(a).

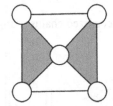

(a) Generalized chain

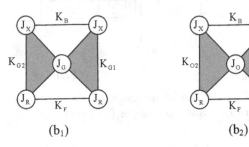

(b) Specialized chains

Figure 7.42   Atlases of generalized kinematic chains and specialized chains (Example 7.6)

Step 3. Specialized chains

In this example, one link is added as a rope. The process of specialization is the same as in Example 7.4. The design requirements and constraints of the rope/belt and fixed joint are as follows:

1. The belt/rope must be a binary link.
2. The belt/rope cannot be adjacent to the frame.
3. Any joint incident to the belt/rope must be a fixed joint.

Here, only the link which is not adjacent to the frame can be assigned to the rope, and two results are obtained as shown in Figure 7.42(b).

Step 4. Specialized chains with particular identities
Following the same representation, the characteristics of all members and joints are assigned in this step except the rope and the fixed joints. Then, the simplest solution, a direct connection, is chosen. As a result, six specialized chains with particular identities are obtained as shown in Figure 7.43.

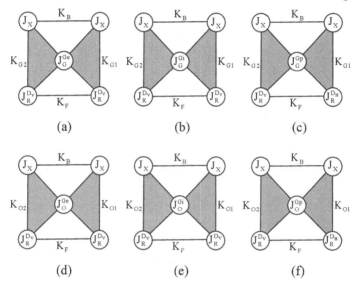

Figure 7.43   Atlas of specialized chains with particular identities – half of the design (Example 7.6)

Step 5. Reconstruction designs
The same two specialized chains with particular identities obtained in the above step along the symmetrical axis are combined to obtain the complete configurations of the designs as shown in Figure 7.44. Note that the rotational direction of the output should be opposite to that of the chariot in the turning motion.

Figure 7.45 shows the corresponding six design concepts for the atlas of specialized chains with particular identities shown in Figure 7.44. The one shown in Figure 7.45(e) is Z. D. Wang's design [33].

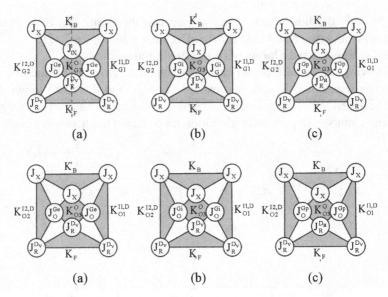

(a)   (b)   (c)

(a)   (b)   (c)

Figure 7.44 Atlas of specialized chains with particular identities (Example 7.6)

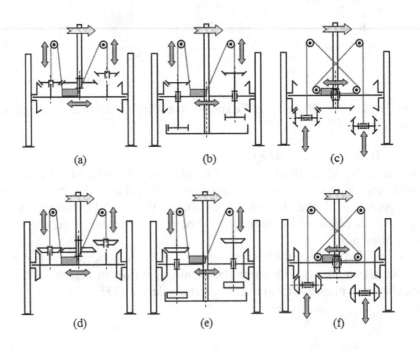

Figure 7.45 Atlas of reconstruction designs (Example 7.6)

References

1. Chen, C.W., Systematic reconstruction design of south pointing chariots (in Chinese), Ph.D. dissertation, Department of Mechanical Engineering, National Cheng Kung University, Tainan, Taiwan, October 2006.
    陳俊瑋，指南車之系統化復原設計，博士論文，國立成功大學機械工程學系，台南，台灣，2006 年 10 月。

2. Yan, H.S. and Chen, C.W., "A systematic approach for the structural synthesis of differential-type south point chariots," JSME International Journal, Series C, Vol. 49, No. 3, pp. 920–929, September 2006.

3. Chen, C.W. and Yan, H.S., "Topological structures of south pointing chariots," Proceedings of the 11th World Congress in Mechanism and Machine Science (IFToMM 2004), Tianjin, China, 1–4 April 2004.

4. Gu Jin Zhu (Notes on the Antiquity and Present Days) (in Chinese) by Chi Bao (Jin Dynasty), Taiwan Commercial Press, Taipei, 1966.
    《古今注》；崔豹[晉朝]撰，台灣商務印書館，台北，1966 年。

5. Zhi Lin (in Chinese) by Yu Xi (Jin Dynasty), Shanghai Wen Yi Publishing House, Shanghai, 1991.
    《志林》；虞喜[晉朝]撰，志林，上海文藝出版社，上海，1991 年。

6. Huang Di Nei Zhuan (in Chinese), author unknown (Southern and Northern Dynasties – Sui and Tang Dynasties), refer to Dao Jiao in China, Chapter 3, Dong Fang Books Publishing Center, Shanghai, 1994.
    《黃帝內傳》；佚名[南北朝-隋唐]，參考：卿希泰編"中國道教"，第三卷，東方出版中心，上海，1994 年。

7. Summary of Zi Zhi Tong Jian (Comprehensive Mirror for Aid in Government) (in Chinese) by Zhu Xi (Song Dynasty), Shanghai Ancient Books Publishing House, Shanghai, 2002.
    《資治通鑑綱目》；朱熹[宋朝]撰，上海古籍出版社，上海，2002 年。

8. Tai Ping Imperial Panorama (in Chinese), edited by Li Fang (Song Dynasty), Taiwan Commercial Press, Taipei, 1983.
    《太平御覽》；李昉[宋朝]等奉敕撰，台灣商務印書館，台北，1983 年。

9. Song Shu (Annals of Rites) (in Chinese) by Shen Yue (Liang Dynasty), Taiwan Commercial Press, Taipei, 1983.
    《宋書》；沈約[梁] 撰，台灣商務印書館，台北，1983 年。

10. Xi Jing Za Ji (in Chinese) by Ge Hong (Jin Dynasty), Yi Wen Publisher, Taipei, 1970.
    《西京雜記》；葛洪[晉朝]撰，藝文出版社，台北，1970 年。

11. Wu Du Fu (in Chinese) by Zuo Si (Jin Dynasty), collected in Zhao Min Wen Xuan by Xiao Tong (Liang Dynasty), Culture Books Publishing House, Taipei, 1977.
    《吳都賦》；左思[晉朝]撰，收於蕭統[梁]編，昭明文選，文化圖書出版社，台北，1977 年。

12. Wei Lue (in Chinese) by Yu Huan (Wei Dynasty), Ding Wen Publishing House, Taipei, 1981.
《魏略》；魚豢[魏]撰，魏略輯本，鼎文出版社，台北，1981年。

13. History of the Three Kingdoms (in Chinese) by Chen Shou (Jin Dynasty), Taiwan Commercial Press, Taipei, 1968.
《三國志》；陳壽[晉朝]撰，台灣商務印書館，台北，1968年。

14. Jin Shu (in Chinese) by Fang Xuan-ling (Tang Dynasty), Taiwan Commercial Press, Taipei, 1983.
《晉書》；房玄齡[唐朝]撰，台灣商務印書館，台北，1983年。

15. Nan Qi Shu (in Chinese) by Xiao Zi-xian (Liang Dynasty), Hong Ye Books, Taipei 1972.
《南齊書》；蕭子顯[梁]撰，宏業書局，台北，1972年。

16. Nan Shi (in Chinese) by Li Yan-shou (Tang Dynasty), World Books, Taipei, 1986.
《南史》；李延壽[唐朝]撰，世界書局，台北，1986年。

17. Jiu Tang Shu (in Chinese) by Liu Xu (Eastern Jin Dynasty), Ding Wen Publishing House, Taipei, 1976.
《舊唐書》；劉昫[東晉]撰，鼎文出版社，台北，1976年。

18. Song Shi (History of the Song Dynasty) (in Chinese) by Tuo Tuo (Yuan Dynasty), Vol. 48, Ding Wen Publishing House, Taipei.
《宋史》；脫脫[元朝]撰，卷四十八，鼎文出版社，台北，1955年。

19. Wei Shu (in Chinese) by Wei Shou (Southern and Northern Dynasties/Northern Qi), Ding Wen Publishing House, Taipei, 1975.
《魏書》；魏收[南北朝/北齊]撰，鼎文出版社，台北，1975年。

20. Shi Shu (in Chinese) by Wei Zheng (Tang Dynasty), Taiwan Commercial Press, Taipei, 1983.
《隋書》；魏徵[唐朝]撰，隋書，台灣商務印書館，台北，1983。

21. New Revised Jiu Tang Shu (in Chinese) by Liu Xu (Eastern Jin Dynasty), Ding Wen Publisher, Taipei, 1976.
《新校本舊唐書》；劉昫[東晉]撰，舊唐書，鼎文出版社，台北，1976年。

22. New Tang Shu (in Chinese) by Ou Yang-xiu (Song Dynasty), Ding Wen Publishing House, Taipei, 1976.
《新唐書》；歐陽修[宋朝]撰，台灣商務印書館，台北，1976年。

23. Yu Hai (in Chinese) by Wang Ying-lin (Song Dynasty), Hua Wen Publishing House, Taipei, 1964.
《玉海》；王應麟[宋朝]撰，華文出版社，台北，1964年。

24. Jin Shi (in Chinese) by Tuo Ke Tuo (Yuan Dynasty), Taiwan Commercial Press, Taipei, 1983.
《金史》；托克托[元朝]撰，台灣商務印書館，台北，1983年。

25. San Cai Tu Hui (in Chinese) by Wang Qi (Ming Dynasty), Zhuang Yan Culture Co., Tainan, Taiwan, 1995.
《三才圖會》；王圻[明朝]撰，莊嚴文化事業公司，台南，台灣，1995年。

26. Gaubil, A., Observations mathèmatiques, astro., geogr., chronol., et phys., tires des anciens livres chinois, Paris, pp. 94–95, 1732.

27. Klaproth, J., Lettre á Humboldt sur l'invention de la boussole, Paris, p. 93, 1834.

28. Hirth, F., "Origin of the Mariner's Compass in China," The ancient history of China, Columbia University Press, New York, pp. 129–130, 1908.

29. Giles, H.A., "The mariner's compass," Adversaria Sinica, No. 7, Shanghai, p. 219, 1909.

30. Moule, A.C., "Textual research on the manufacture of Yan Su's and Wu De-ren's south pointing chariots from the Song Dynasty," translated by Zhang Yin-lin, Qinghua Journal, Beijing, Vol. 2, pp. 457–467, 1925.

31. Hashimoto M., "Origin of the compass," Memoir's of the Research Department of the Toyo Bunko (The Oriental Library), Tokyo, No. 1, pp. 67–92, 1926.

32. Mikami Y., "The chou-jen-chuan of yuan yuan," Isis, Chicago, Vol. II, p. 124, 1928.

33. Wang, Z.D., "Investigations and reproduction in model form of the south pointing chariot and the hodometer (in Chinese)," Beijing Academy of Sciences, Historical Journal, Beijing, No. 3, pp. 1–47, 1937.
    王振鐸，"指南車記里鼓車之考證與模製"，史學集刊，科學出版社，第 3 期，北京，第 1–47 頁，1937 年。

34. Lanchester, G., "The Yellow Emperor's south pointing chariot," a speech script at the China Society of Britain, London, 1947.

35. Needham, J., Science and civilization in China (in Chinese), Vol. 4, Taiwan Commercial Press, Taipei, pp. 286–303, 1965.
    李約瑟，中國科學與文明，第 4 冊，台灣商務印書館，台北，第 286–303 頁，1965 年。

36. Li, S.H., The south pointing carriage and the mariner's compass (in Chinese), Yee Wen Publishing House, Taipei, 1959.
    李書華，指南車與指南針，藝文印書館，台北，第 12–19 頁，1959 年。

37. Lu, J.Y., "Summary of south pointing chariot study (in Chinese)," History Monthly, Taipei, No. 80, pp. 80–84, 1994.
    陸敬嚴，"指南車研究概述"，歷史月刊，台北，第 80 期，第 80–84 頁，1994 年。

38. Lu, Z.M., "An analysis of the ancient Chinese south pointing chariot (in Chinese)," Journal of Sichuan University, Sichuan, No. 2, pp. 95–101, 1979.
    盧志明，"中國古代指南車的分析"，西南大學學報，四川，第 2 期，第 95–101 頁，1979 年。

39. Yan, Z.R., "Principles and structures of south pointing chariot in ancient China," Journal of Shanghai Mechanical College (in Chinese)," Shanghai, No. 1, pp. 31–41, 1984.
    顏志仁，"中國古代指南車的原理與構造"，上海機械學院學報，上海，第 1 期，第 31–40 頁，1984 年。

40. Yan, Z.R., "The south pointing chariot," Middle School Science and Technology (in Chinese)," Shanghai, No. 5, pp. 32–33, 1982.
    顏志仁，"指南車"，中等學校科學與技術，上海，第 5 期，第 32–33 頁，1982 年。

41. Yang, Y.Z., "Design of south pointing chariot mechanisms (in Chinese)," Mechanical Engineering, Taipei, No. 154, pp. 18–24, 1986.
    楊衍宗，"指南車機構設計"，機械工程，台北，第 154 期，第 18–24 頁，1986 年。

42. Muneharu, M. and Satoshi, K., "Study of the mechanics of the south pointing chariot – the south pointing chariot with the bevel gear type differential gear train," Transactions of Japan Society of Mechanical Engineering, Part C, Vol. 56, pp. 462–466, 1990.

43. Muneharu, M. and Satoshi, K., "Study of the mechanics of the south pointing chariot – 2nd report, the south pointing chariot with the external spur-gear-type differential gear train," Transactions of Japan Society of Mechanical Engineering, Part C, Vol. 56, pp. 1542–1547, 1990.

44. Santander, M., "The Chinese south-seeking chariot," American Journal of Physics, pp. 782–790, 1992.

45. Hsieh, L.C., Jen, J.Y., and Hsu, M.H., "Systematic method for the synthesis of south pointing chariot with planetary gear trains," Transactions of Canadian Society for Mechanical Engineering, Vol. 20, pp. 421–435, 1996.

46. Chen, Y.J., "A south pointing chariot with frictional transmission," Taiwan (ROC) Patent, No. 371043, 1999.
    陳英俊，摩擦傳動指南車，中華民國新型專利第 371043 號，1999 年。

47. Liu, X.Z., "Chinese inventions in power transmission (in Chinese)," Qinghua Engineering Reports, Vol. 2, pp. 40–47, 1954.
    劉仙洲，"中國在傳動機件方面的發明"，清華學報，北京，第 2 卷，第 40–47 頁，1954 年。

48. Liu, X.Z., A history of Chinese engineering inventions (in Chinese), Science Press, Beijing, Vol. 1, 1962.
    劉仙洲，中國機械工程發明史-第一編，科學出版社，北京，1962 年。

49. Sleeswyk, A.W., "Reconstruction of the south pointing chariots of the Northern Song Dynasty, escapement and differential gearing in 11th century China," Chinese Science, Pennsylvania, pp. 4–36, 1977.

50. Yan, Z.R., "Circling infinitely, yet the driving method remains the same (in Chinese)," Popular Machinery, Shanghai, No. 1, pp. 18–19, 1983.
    顏志仁，"運轉不窮而司方如一"，大眾機械，上海，第 1 期，第 18–19 頁，1983 年。

51. Bagci, C., "Elementary theory for the synthesis of constant direction pointing chariots," Gear Technology, Vol. 5, pp. 31–35, 1988.

52. Yan, H.S., Creative Design of Mechanical Devices, Springer, Singapore, 1998.

Chapter 8 Walking Machines

Studies and publications on modern walking machines appeared only in the last 100 years. However, according to ancient Chinese records, walking machines, devices that mimic the horse or ox using mechanical legs, might have been created before the time of Christ, especially the Wooden Horse Carriage (木車馬) of Lu Ban (魯般) around 480 BC and the Wooden Ox and Gliding Horse (木牛流馬) of Zhu-ge Liang (諸葛亮) around 230 AC. These inventions were treated as novelties, and they can be found in literary records but without surviving hardware.

This chapter systematically reconstructs all feasible designs of Lu Ban's wooden horse carriage with mechanical legs that meet the scientific and technological standards of the subject's time period [1–6]. Historical background, literature and development of ancient Chinese walking machines are studied and discussed. Works involving restoration are also presented. Finally, two examples based on different design requirements and constraints are provided.

8.1 Lu Ban the Man

Lu Ban (魯般) (~507–444 BC), whose original family name was Gong-shu (公輸), was a master carpenter and inventor in the State of Lu during the Era of Spring and Autumn (770–481 BC). Although believed to be a native of Dunhuang in ancient China, there is no definite reference to his ancestral origin [7].

According to legend, Lu Ban was not fond of learning when he was young, but was enlightened when he studied under the scholar Zi Xia (子夏). Later, by coincidence, he received lessons from Bao Lao-dong (鮑老董), and learned fine wood-carving skills. Because the family of Lu Ban was engaged in handicrafts manufacturing, his major profession became carpentry, specializing in building houses and making furniture and other utilities.

Since the working tools and devices were primitive during his time, work had to be done manually. Each job was drudgery. But Lu Ban had a muscular physique; he was bright, and his works were refined. The houses he built were strong and durable, and the furniture he made was elegant. People near and far came to him for carpentry work, and he was kept busy all the time. However, even if business was good, if he could not produce on time, it would still be problematic. Lu Ban therefore began inventing and improving his tools, including the saw, planning tool, drill and shovel, carpenter's ink marker, carpenter's square, and hook and peg on the ink marker. Furthermore, Lu Ban's wife was also believed to be an inventor. She made an umbrella for Lu Ban to carry, enabling him to work outside under any weather condition.

Lu Ban was not only an excellent carpenter; he was also an outstanding inventor. He invented the bridge-tower which was a very tall contraption used to attack castles for the State of Chu. He also made a wooden kite to spy on enemy castles. He manufactured arms for Chu's warships, and invented a kind of hook and iron mast that enabled Chu's ships to maneuver efficiently upstream. But what fascinated people most were his designs for the wooden kite and the wooden horse carriage.

8.2 Literary Works Related to the Wooden Horse Carriage

The following are literary works about Lu Ban found in ancient Chinese historical records.

Wooden kite (木鳶)

1. Mo Zi · Chapter 5 · Article Lu Wen 《墨子·第五卷魯問篇》 [8]

    Lu Ban was bragging about how amazing his flying kite was, made from bamboo sticks, because it could fly in the sky for three days. Mo Zi reminded him that his flying kite was no match to the cart made by the artisan, because it could carry 50 dan (a unit of weight of dry measure for grain) with only a piece of wood three cun thick. 『公輸子削竹木以爲鵲，成而飛之三日不下，公輸子自以爲至巧。子墨子謂公輸子曰，子之爲鵲也不如翟之爲車轄，須臾劉三寸之木，而任五十石之重。』

2. Huai Nan Zi · Chapter 11 · Article Qi Su Xun 《淮南子·第十一卷齊俗訓》 [9]

    Mo Zi told Lu Ban that his kite that could fly for three days in the sky could not be put to better use. 『魯班墨子以木爲鳶而飛之三日，不集而不可使爲工也。』

3. Lun Heng · Chapter 8 · Article Ru Zeng 《論衡·第八卷儒增篇》 [10]

According to book Ru Shu, Lu Ban and Mo Zi built a wooden kite that could fly for three days. I believe this is possible, because if his wooden horse carriage could travel for three days, and he applied the same principle with the wooden kite, then the kite could fly for three days, too. 『儒書稱魯班墨子之巧刻木爲鳶，飛之三日而不集。夫言其以木爲鳶飛之可也，言其三日不集增之也，夫刻木爲鳶以象鳶形安能飛而不集乎既能飛翔，妥能至於三日如審有機關一飛遂翔不可復下，則當言遂飛不當言三日。』

4. Lun Heng・Chapter 16, article Luan Long《論衡・第十六卷亂龍篇》 [10]

    Lu Ban and Mo Zi built the wooden kite which could fly for three days, it was an amazing creation." 『魯班墨子刻木爲鳶飛之三日而不集，爲之巧也。』

Wooden horse carriage (木車馬)

The most realistic account of the wooden kite was in the book Ru Shu 《儒書》 [11]. Mo Zi (墨子) was born in 468 BC, 39 years later than Lu Ban. According to legend, Mo Zi was skilled in inventing tools. He spent 3 years building a wooden bird, but it could only stay in the sky for 1 day. Lu Ban bragged that his wooden bird could stay for 3 days, and that his work was superior to that of Mo Zi. Mo Zi later commented that the wooden bird of Lu Ban was no match for the wooden carts built by artisans, for these carts could carry 50 dan of heavy load with only a piece of wood three cun thick.

The book Mo Zi 《墨子》 [8] was jointly written by Mo Zi and his students. The discussion between Mo Zi and Lu Ban, as documented in Chapter 5 of the book, would therefore be realistic. In addition, there were also records of the wooden bird of Lu Ban in Chapter 11 of the book Han Zi Yu Ping 《韓子迂評》 [12], Chapter 11 of the book Huai Nan Zi 《淮南子》 [9], and Chapters 8 and 11 of the book Lun Heng 《論衡》 [10].

According to legend, Lu Ban was a filial son. He built a wooden horse carriage for his aged mother so that she would not tire herself when she went out. The carriage was designed to move without need of human control. It first appeared in the book Lun Heng by Wang Chong (王充, ~ 27–97 AC) in the Eastern Han Dynasty [10]. It states: "It is said that Lu Ban was mourning of the loss of this mother. He built a wooden horse carriage which was well equipped and needed no manual intervention. When his mother rode on it, it sped away never to return. If the mechanical principle used on the wooden kite was the same as that used in the wooden horse carriage, then it would fly in the sky. If the mechanism he installed on the carriage would not operate automatically for more than three days, then he would have found his mother along the three-day route." 『猶世傳言曰：「魯班巧，亡其母也。」言巧工爲母作木車馬，木人御者，機關備具，載母其上，一驅不還，遂失其母。如木鳶機關備具與木車馬等則逐飛不集，機關爲須臾間不能遠過三日，則木車等亦宜三日止於道路，無爲徑去以失其母。』 (Figure 8.1)

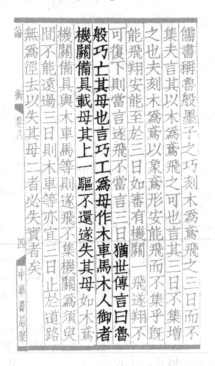

Figure 8.1 Description of Lu Ban's wooden horse carriage in Lun Heng《論衡》[10]

The work of Wang Chong was primarily a response to the book Ru Shu. There was a part in it that questioned the credibility of the claim that the flying contraption of Lu Ban could stay in the sky for 3 days. Wang Chong believed, however, that if the carriage of Lu Ban could move automatically without stopping, then if Lu Ban had applied the same mechanical principle for his kite, then it could also fly for 3 days without falling. Furthermore, if the wooden carriage could not move on its own, then when Lu Ban's mother was riding in the carriage, the carriage should have stopped moving somewhere, enabling Lu Ban to find his mother along the 3-day carriage route. According to historical records, however, Lu Ban's mother was never found.

Lu Ban lived in Dunhuang, a place full of mountainous slopes. This may also strongly imply that his carriage could move in the rugged mountainous terrain based on the principle of inertia and using the concept of balance of energy.

Literary works about the carriage were few, but there were many records of Lu Ban's wooden bird in history books. Therefore, based on the book Lun Heng by Wang Chong, the existence of the wooden horse carriage of Lu Ban could be proven indirectly. If the wooden bird had existed, then the carriage must have existed also because the design of the flying device should be more difficult than the ground carriage. In addition, if the carriage were operated by linkage mechanisms, it would not be a problem for a carpenter like Lu Ban. The most critical issues were the correct dimensions and assembly of the parts. To somebody like Lu Ban who had no modern technological background, the structure of the device would certainly be based on experiments done with rich engineering experiences. Therefore, the creation of the wooden horse carriage was possible.

The wooden horse carriage of Lu Ban was invented under such a condition; but the invention was treated as a novelty and quickly disappeared. No relevant information about his invention was recorded because the scholars during that time were ignorant of technological knowledge. Nevertheless, it is the earliest story of ancient Chinese walking machine.

Based on the records of the above-mentioned history books, the existence of the wooden horse carriage and the wooden bird should be valid. Seen from a modern technological perspective, if the account of Wang Chong was true, then the wooden horse carriage would be one of the major inventions of ancient China. Unfortunately, the invention was lost during the Eastern Han Dynasty (AD 25–220). Consequently, people are not able to see this magnificent work today.

In past history, very few scholars studied lost ancient Chinese walking machines. However, in recent decades, scholars who believe that the wooden horse carriage of Lu Ban was an enigmatic ancient invention have been reproducing the device. Around 1986, Wang Jian (王涧) of Urumqi in province Xinjiang of China, built a wooden horse carriage based on his ingenious experience and sense of practicality, Figure 8.2 [5, 6]. This design is composed of a walking mechanism with leg function and a trailer with balance function. The walking mechanism has four sets of eight-bar linkage with the same configurations.

Figure 8.2 The wooden horse carriage by Wang Jian

8.3 Literary Works Related to the Wooden Ox and Gliding Horse

Another famous walking machine named wooden ox and gliding horse (木牛流馬) in ancient China was invented by Zhu-ge Liang (諸葛亮) in the Era of Three Kingdoms (AD 220–280).

The book History of the Three Kingdoms 《三國志》 by Chen Shou (陳壽) recorded that [13]: "In the ninth years, Zhu-ge Liang staged another war at mountain Qi, used the wooden ox to transport supplies. The army retreated when the supply was exhausted. The general of Wei, Zhang He, was killed in a battle by the army of Shu. In spring of the twelfth year, Liang led his army through the valley, using the wooden ox as transport. His army camped at Wu Zhang Yuan, opposing the army of King Si-ma Xuan in Wei Nan."『九年，亮復出祈山，以木牛運，糧盡退軍，與魏將張郃交戰，射殺郃。十二年春，亮率大眾由斜谷出，以木牛運，據武功五丈原，與司馬宣王對於渭南。』This narration clearly indicated that the cattle machine was created for food supplies in the rough terrain of mountainous regions. It first appeared in AD 209, and was a machine invented due to the necessities of war.

Since the legendary cattle machine of Zhu-ge Liang did not survive to the present and historical records on the subject are not comprehensive, scholars have different opinions about its existence. Several theories have been proposed in the past hundreds of years. Two major credible theories are: the device is a single-wheel barrow (Figure 8.3), and the device is a four-legged walking contraption.

Figure 8.3 A single-wheel barrow [14]

Many books and records supported the theory that the wooden ox and gliding horse was a four-legged walking machine. The earliest one was in the Biography of Zu Chong-zhi in the book Nan Qi Shu《南齊書·祖沖之傳》[15]. It stated that: "Zhu-ge Liang invented machines, namely the wooden ox and gliding horse, that were powered neither by the wind nor water, and the wooden ox and gliding horse carried loads without need of man's physical efforts."『以諸葛亮有木牛流馬,乃造一器,不因風水,失機自運,不勞人力。』The book further described that when left on a slope, the wooden ox and gliding horse moved down without human intervention.

Author Li Fang (李昉), in his Tai Ping Imperial Panorama《太平御覽》[16], further classified the wooden ox and gliding horse as "skilled invention" rather than "military gadget," adding support to the theory. The Instructions on Making Wooden Ox and Gliding Horse in book Zhu-ge Liang Collection《諸葛亮集·作木牛流馬法》[17] described the wooden ox and gliding horse as raising two shafts to walk four hooves forward, while a man moves six chi. Furthermore, there are many impartial historical narratives on the wooden ox and gliding horse, such as: Biography of Zhu-ge Liang《諸葛亮傳》[18], Yuan He Prefecture and County Policies 《元和郡縣志》 [19], Tale of the End of the Han Dynasty《漢末傳》[20], Other Tales of Pu Yuan《蒲元別傳》[21], Collective Analysis on the History of the Three Kingdoms《三國志集解》[22], Book of Shu 《蜀書》[23], History of the Political Units of Han Dynasty《漢郡國志》[24], Wei's Records of the Spring and Autumn Period《魏氏春秋》[25], and Inscriptions in the Zhu-ge Temple《諸葛武侯廟碑銘》[26]. Materials that are helpful in the reproduction of the cattle machine include: Zi Zhi Chronicle《資治通鑑》[27], Book of General References《通典》[28], Romance of the Three Kingdoms《三國演義》[29], and Instructions on Making Wooden Ox and Gliding Horse in book Zhu-ge Liang Collection [17].

Although literature did not specifically record the use of the wooden ox and gliding horse, the importance and contributions of the device to the Shu Kingdom in the Era of Three Kingdoms (AD 220–280) can be rationalized from their contents.

The book Zhu-ge Liang Collection – Instructions on Making Wooden Ox and Gliding Horse described this machine as *"being able to carry heavy loads but advances slowly, suitable for cumbersome tasks and not for light applications."*『載多而行少，宜可大用，不可小使。』 This statement directly pointed out the main function of the device. There are two theories regarding its description [30, 31]:

1. The cattle machine was suited for military applications and not for transporting light cargoes. This theory assumes that the device was created for transport purposes during war. The production of such devices was not easy and high technology was involved. Thus, it seems impractical for this device to be used as a civilian transport.

2. The cattle machine could carry heavy loads, but it moved slowly. The conditions during that time, however, did not allow it to make many trips.

If the reference material is understood in its entirety, the second theory seems more rational.

After the wooden ox and gliding horse of Zhu-Ge Liang was lost to civilization, Zu Chong-zhi (祖沖之, AD 429–500) reproduced his version of the machine, but it too was lost.

Around 1986, Wang Jian of Urumqi in province Xinjiang in China, built his version of the wooden ox and gliding horse based on his ingenious experience and sense of practicality, Figure 8.4. This design is a 33-link, four-legged walking machine. The mechanisms on both sides of the device are symmetrical, and they are connected to a common crank that causes the links of the two sides to move in opposite directions. When the legs on one side of the machine move during the supporting stage, they initiate striding movement of the two legs on the other side. The four links of the front hooves correspond with the four links of the rear hooves; and they are connected to two equal-length long links. When the rear hoof moves, the momentum is passed through the long links, forcing the front hoof to move in step. The knee joint of the front leg is in front of the point where the hip joint touches the ground. When the front leg is raised, the joint protrudes forward. On the contrary, the knee joint of the rear leg is behind the point where the hip joint touches the ground. When the leg is raised, the knee joint protrudes backward. In this way, the movement of the machine closely approximates that of real cattle. And, the two legs on either side of the machine operate in a synchronous manner when it is moving.

8.4   Other Walking Machines

In the Song Dynasty (AD 960–1279), the interpretations regarding the wooden ox and gliding horse became diverse. However, it is noteworthy that the poem Yang Ma Ge (秧馬歌) by Su Shi (蘇軾) (AD 1037–1101) described a horse planting rice seedlings in the field. The reasonable conjecture should be that there is a horse-shaped mechanical device helping the job of a farmer. Furthermore, the works of Yuan Mei (袁枚) in the Qing Dynasty (AD 1644–1911) mentioned that Jiang Yong (江永, AD 1681–1762) could invent magical walking devices. For example, Chapter 107 in the Bei Zhuan Collection《碑傳集》said that: "Someone rode a wooden donkey that eats nothing and talks silence outside the city, and people took it as a weird thing. However, this amazed device was functioned mechanically based on Zhu-ge Liang's approach, not a weird thing."『行城外騎一木驢，不食不鳴，人以爲妖。笑曰：此武侯成法，不過用機耳，非妖也。』[5, 6]

Figure 8.4 The wooden ox and gliding horse by Wang Jian

8.5   Reconstruction Design

Subject to limited historical records and technological constraints in ancient eras, the procedure for the reconstruction of design concepts of possible wooden horse carriages is shown in Figure 8.5. It consists of the following four steps:

Step 1.   Design specifications

Since no original designs are available from historical archives, and based on exhaustive literature study, basic design specifications regarding the topological structure of the wooden horse carriage are defined as follow:

1. It is a quadruped walking machine that generates specific gait locomotion, and it mimics the motion of a real horse.
2. Each leg mechanism is a planar linkage with simple revolute joints and one degree of freedom.
3. A carriage is attached to the body of the wooden horse to provide the function of balance.

Step 2. Generalized kinematic chains
The second step is to obtain or identify the atlas of generalized kinematic chains with the required numbers of links and joints subject to defined design specifications (topological characteristics by applying the algorithm of number synthesis [32] or simply identified from Section 4.4.

Step 3. Specialized chains
The third step is to have the atlas of specialized chains with assigned types of links and joints subject to the concluded design requirements and constraints for each generalized kinematic chain obtained in Step 2.

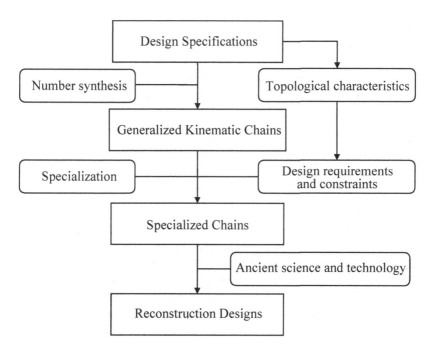

Figure 8.5 Process of reconstruction design

Step 4. Reconstruction designs

The last step is to obtain the atlas of reconstruction designs from the atlas of specialized chains according to the motion and function requirements of the ancient machinery, and by utilizing the mechanical evolution and variation theory to perform a mechanism equivalent transformation. Ancient science theories and technologies of the subject's time period are applied to find appropriate and feasible mechanisms that can be considered as the reconstruction designs.

8.6 Design Examples

Here, the target wooden horse carriage comprises a four-legged walking machine with identical leg mechanisms and a trailer. All feasible topological structures of the leg mechanisms of six-bar (Example 8.1) and eight-bar (Example 8.2) types are synthesized in the following two examples [1–4].

[Example 8.1]
Six-bar type wooden horse carriages.

For a planar six-bar leg mechanism with simple revolute joints and one degree of freedom, based on Equation (2.1) for the degrees of freedom $F_p = 1$, the number of members $N_L = 6$, and the number of degrees of constraint of a revolute joint $C_{pR} = 2$,

$$
\begin{aligned}
N_{JR} &= [3(N_L - 1) - F_p]/C_{pR} \\
&= [(3)(6 - 1) - 1]/2 \\
&= 14/2 \\
&= 7
\end{aligned}
$$

the number of joints (N_{JR}) is seven.

Therefore, the leg mechanism of six-bar type wooden horse carriages should be a (6, 7) kinematic chain. And, there are two (6, 7) kinematic chains shown in Figure 4.22 and again in Figure 8.6.

Once the atlas of the kinematic chains is obtained, all possible specialized chains can be identified through the following substeps:

1. For each kinematic chain, identify the thigh link (member 3, K_{Lt}) and the shank link (member 4, K_{Ls}) that are adjacent to each other for all possible cases.
2. For each case obtained in substep 1, identify the ground link (member 1, K_F).
3. For each case obtained in substep 2, identify the crank (member 2, K_{Lc}).

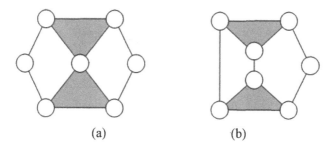

(a)   (b)

Figure 8.6 Atlas of (6, 7) kinematic chains (Example 8.1)

These substeps are carried out subject to the following design requirements and constraints:
1. It has a ground link (frame) as the body.
2. It has a crank.
3. The crank of the leg mechanism is adjacent to the body, and the fixed pivots of all the four cranks are coaxial.
4. It has a thigh link.
5. It has a shank link.
6. The crank, the thigh link, and the shank link must be distinct members.
7. The crank cannot be adjacent to the thigh link or the shank link.
8. The thigh link is adjacent to the body and the shank link.
9. The shank link cannot be adjacent to the body, but is adjacent to the thigh link. In addition, there is a foot point (coupler point) on the shank link to generate a path curve and to contact the ground.

In what follows, the kinematic chain shown in Figure 8.6(a) is chosen as an example for the process of specialization.

Thigh link (member 3, K_{Lt}) *and shank link* (member 4, K_{Ls})
Since there must be a thigh link which is adjacent to the shank link, four specialized chains with identified thigh link and shank link are available as shown in Figure 8.7.

Ground link (member 1, K_F)
Since there must be a link as the frame, the ground link can be identified based on the design requirements and constraints as follows:
1. For the case shown in Figure 8.7(a), the assignment of the ground link generates three results, Figure 8.8(a)–(c).
2. For the case shown in Figure 8.7(b), the assignment of the ground link generates two results, Figure 8.8(d) and (e).
3. For the case shown in Figure 8.7(c), the assignment of the ground link generates three results, Figure 8.8(f)–(h).

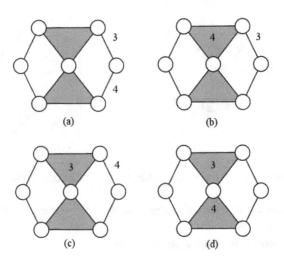

Figure 8.7 Specialized chains with identified thigh link and shank link (Example 8.1)

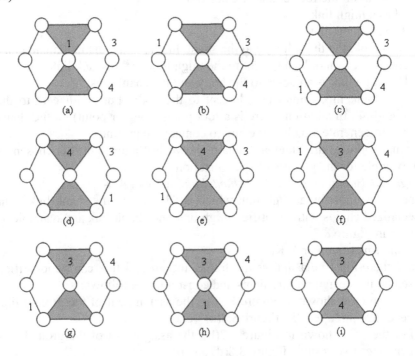

Figure 8.8 Specialized chains with identified thigh link, shank link, and ground link (Example 8.1)

4. For the case shown in Figure 8.7(d), the assignment of the ground link generates one nonisomorphic result, Figure 8.8(i).

Therefore, nine specialized chains with identified thigh link, shank link, and ground link are available, as shown in Figure 8.8.

Crank (member 2, K_{Lc})

Since there must be a crank that is adjacent to the ground link and that cannot be adjacent to the thigh link or the shank link, the crank can be identified as follows:

1. For the case shown in Figure 8.8(a), the assignment of the crank generates two results, Figure 8.9(a_1) and (a_2).
2. For the case shown in Figure 8.8(b), the assignment of the crank generates two results, Figure 8.9(a_3) and (a_4).
3. For the case shown in Figure 8.8(c), the assignment of the crank generates two results, Figure 8.9(a_5) and (a_6).
4. For the case shown in Figure 8.8(d), the assignment of the crank generates one result, Figure 8.9(a_7).
5. For the case shown in Figure 8.89(e), the assignment of the crank generates two results, Figure 8.9(a_8) and (a_9).
6. For the case shown in Figure 8.8(f), the assignment of the crank generates one result, Figure 8.9(a_{10}).
7. For the case shown in Figure 8.8(g), the assignment of the crank generates two results, Figure 8.9(a_{11}) and (a_{12}).
8. For the case shown in Figure 8.8(h), the assignment of the crank generates two results, Figure 8.9(a_{13}) and (a_{14}).
9. For the case shown in Figure 8.8(i), the assignment of the crank generates one result, Figure 8.9(a_{15}).

Therefore, 15 specialized chains with identified thigh link, shank link, ground link, and crank are available, as shown in Figure 8.9(a_1)–(a_{15}).

By following the same process of specialization for the kinematic chain shown in Figure 8.6(b), 17 specialized chains with identified thigh link, shank link, ground link, and crank are available, as shown in Figure 8.9(b_1)–(b_{17}). In summary, there are a total of 32 feasible specialized chains available as the leg mechanism for six-bar type wooden horse carriages as shown in Figure 8.9(a_1)–(b_{17}).

Figure 8.10(a_1)–(b_{17}) show the corresponding schematic formats of the leg mechanisms, providing the atlas of all possible designs. Figure 8.11 shows a physical model of a six-bar type wooden horse carriage based on the concept from Figure 8.10(b_4) [3].

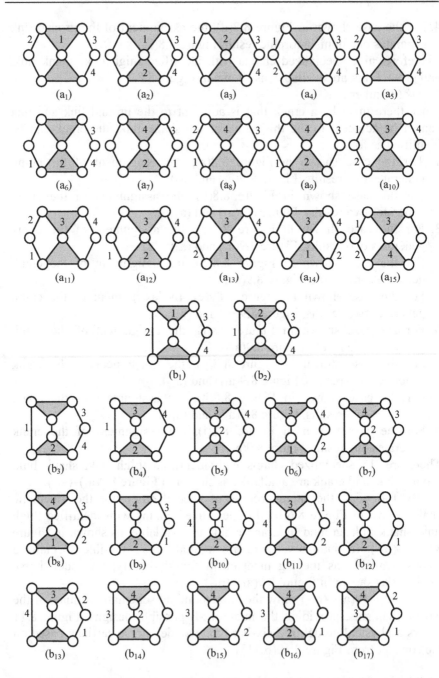

Figure 8.9 Specialized chains with identified thigh link, shank link, ground link, and crank (Example 8.1)

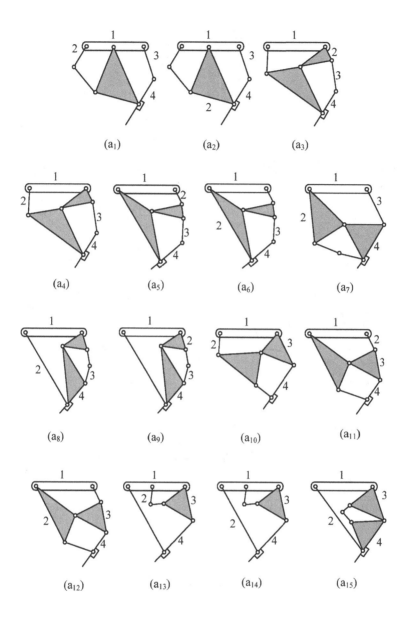

Figure 8.10 Atlas of leg mechanisms for six-bar type wooden horse carriages (Example 8.1)

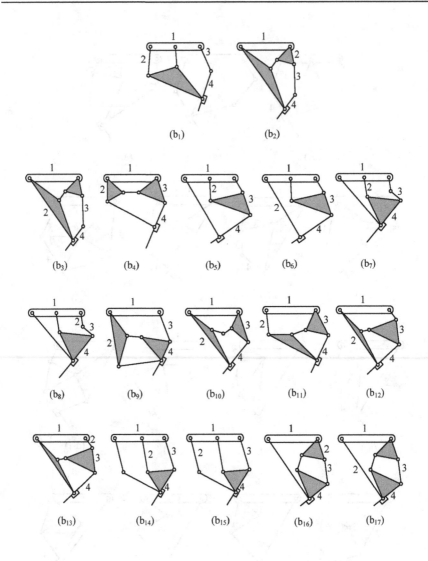

Figure 8.10 (*Continued*)

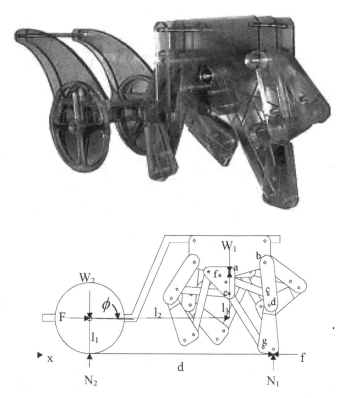

Figure 8.11 Physical model of a six-bar type wooden horse carriage (Example 8.1) [3]

[Example 8.2]
Eight-bar type wooden horse carriages.

For a planar eight-bar leg mechanism with simple revolute joints and one degree of freedom, based on Equation (2.1) for the number of degrees of freedom $F_p = 1$, the number of members $N_L = 8$, and the number of degrees of constraint of a revolute joint $C_{pR} = 2$,

$$
\begin{aligned}
N_{JR} &= [3(N_L - 1) - F_p]/C_{pR} \\
&= [(3)(8 - 1) - 1]/2 \\
&= 20/2 \\
&= 10
\end{aligned}
$$

the number of joints (N_{JR}) is ten. Therefore, the leg mechanism of eight-bar type wooden horse carriages should be an (8, 10) kinematic chain. And, there are 16 (8, 10) kinematic chains as shown in Figure 4.23 and again here in Figure 8.12.

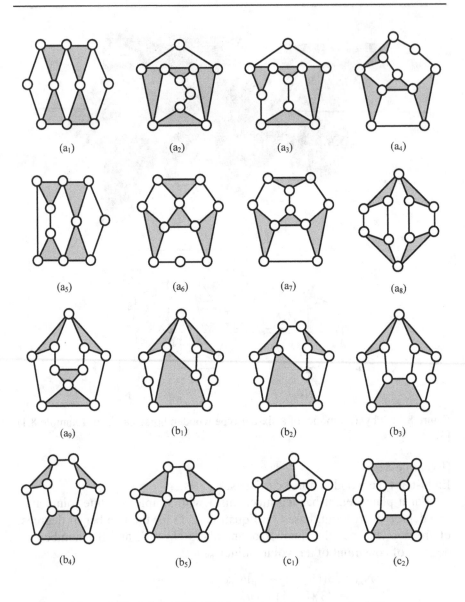

Figure 8.12 Atlas of (8, 10) kinematic chains (Example 8.2)

Once the atlas of the kinematic chains is obtained, all possible specialized chains can be identified through the following substeps:

1. For each kinematic chain, identify the ground link (member 1, K_F) for all possible cases.
2. For each case obtained in substep 1, identify the thigh link (K_{Lt}).
3. For each case obtained in substep 2, identify the shank link (K_{Ls}).
4. For each case obtained in substep 3, identify the crank (K_{Lc}).

These steps are carried out subject to the following design requirements and constraints:

1. It has a ground link as the body, and the ground link is a multiple link.
2. It has a crank, and the crank is a binary link.
3. The crank of the leg mechanism is adjacent to the body, and the fixed pivots of all the four cranks are coaxial.
4. It has a thigh link.
5. It has a shank link.
6. The crank, the thigh link, and the shank link must be distinct members.
7. The crank cannot be adjacent to the thigh link or the shank link.
8. The thigh link is adjacent to the body and the shank link.
9. The shank link cannot be adjacent to the body, but is adjacent to the thigh link. There is also a foot point (coupler point) on the shank link to generate a path curve and to contact the ground.

In what follows, the kinematic chain shown in Figure 8.12(a) is chosen as an example for the process of specialization.

Ground link (K_F)

Since there must be a multiple link as the ground link, and due to the symmetry of links, anyone of the four ternary links in Figure 8.12(a) can be identified as the ground link.

Thigh link (K_{Lt})

Since there must be a thigh link and the thigh link must be adjacent to the ground link, those three links adjacent to the ground link can be assigned as the thigh, Figure 8.13(a)–(c).

Shank link (K_{Ls})

Since there must be a shank link and the shank link must be adjacent to the thigh link, the shank link can be identified as follows:

1. For the case shown in Figure 8.13(a), the assignment of the shank link generates two results, Figure 8.14(a) and (b).
2. For the case shown in Figure 8.13(b), the assignment of the shank link generates two results, Figure 8.14(c) and (d).
3. For the case shown in Figure 8.13(c), the assignment of the shank link generates one result, Figure 8.14(e).

Therefore, five specialized chains with identified ground link, thigh link, and shank link are available as shown in Figure 8.14(a)–(e).

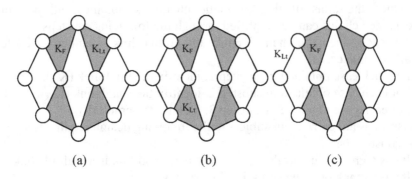

(a)   (b)   (c)

Figure 8.13 Specialized chains with identified ground link and thigh link (Example 8.2)

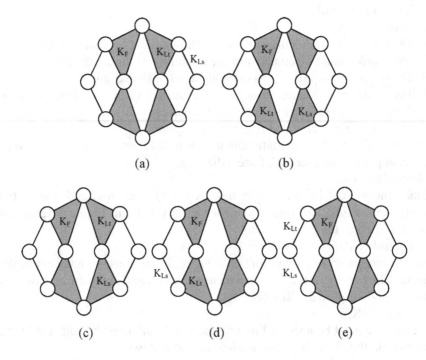

(a)   (b)

(c)   (d)   (e)

Figure 8.14 Specialized chains with identified ground link, thigh link, and shank link (Example 8.2)

Crank (K$_F$)

Since there must be a binary link as the crank that is adjacent to the ground link, but that cannot be adjacent to the thigh link or the shank link, the crank can be identified as follows:

1. For the case shown in Figure 8.14(a), the assignment of the crank generates one result, Figure 8.15(a).
2. For the case shown in Figure 8.14(b), the assignment of the crank generates one result, Figure 8.15(b).
3. For the case shown in Figure 8.14(c), the assignment of the crank generates one result, Figure 8.15(c).
4. For the case shown in Figure 8.14(d), since the binary link adjacent to the ground link is also adjacent to the shank link, no result can be generated.
5. For the case shown in Figure 8.14(e), since the binary link adjacent to the ground link is already assigned as the thigh link, no binary link is available to be assigned as the crank and no result can be generated.

Therefore, three specialized chains with identified ground link, thigh link, shank link, and crank are available, as shown in Figure 8.15.

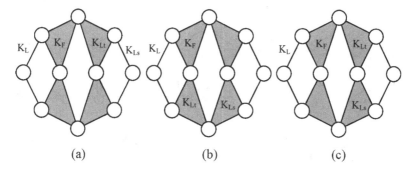

(a)   (b)   (c)

Figure 8.15 Specialized chains with identified thigh link, shank link, ground link, and crank (Example 8.2)

By following the same process of specialization for the kinematic chains shown in Figure 8.12(b)–(p), 114 specialized chains with identified thigh link, shank link, ground link, and crank are available. In summary, there are a total of 117 feasible specialized chains available as the leg mechanism for eight-bar type wooden horse carriages as shown in Figure 8.16(a$_1$)–(p$_3$).

Figure 8.17 shows some of the corresponding schematic format of the leg mechanisms. It is interesting to note that the mechanism shown in Figure 8.17(b) is the leg mechanism of the wooden horse carriage by Wang Jian as shown in Figure 8.2.

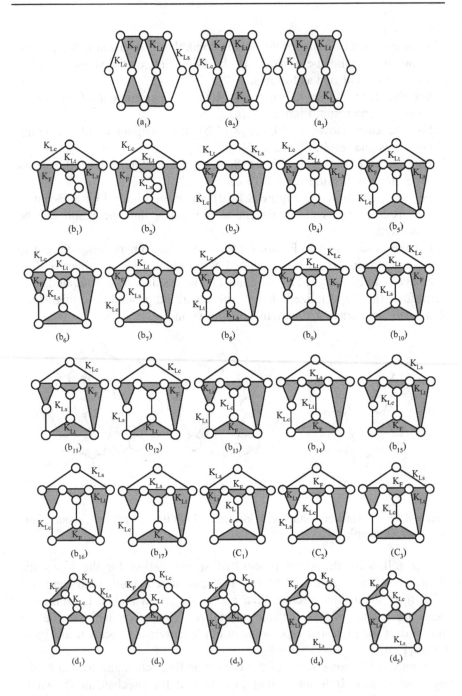

Figure 8.16 Atlas of specialized chains with identified thigh link, shank link, ground link, and crank (Example 8.2)

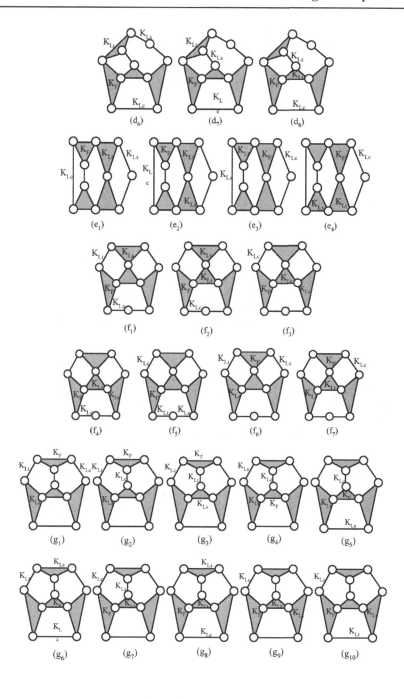

Figure 8.16 (*Continued*)

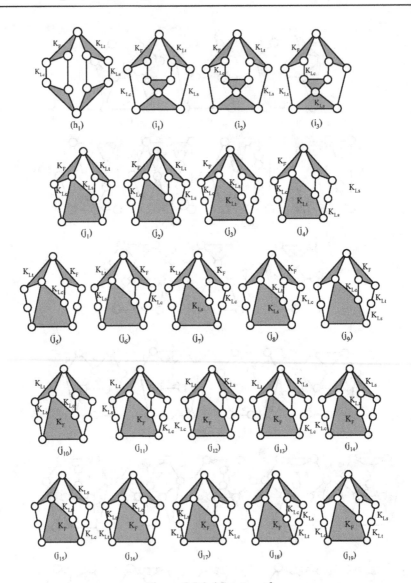

Figure 8.16 (*Continued*)

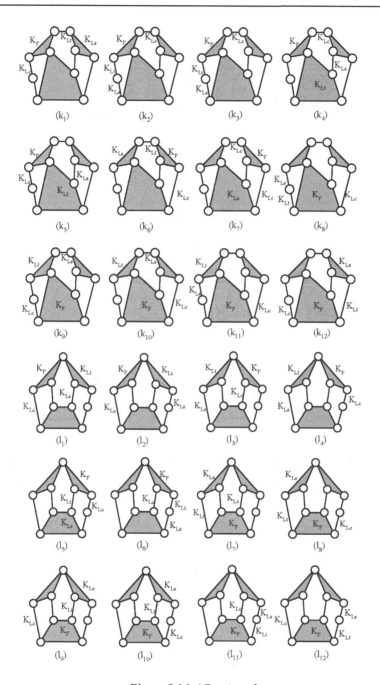

Figure 8.16 (*Continued*)

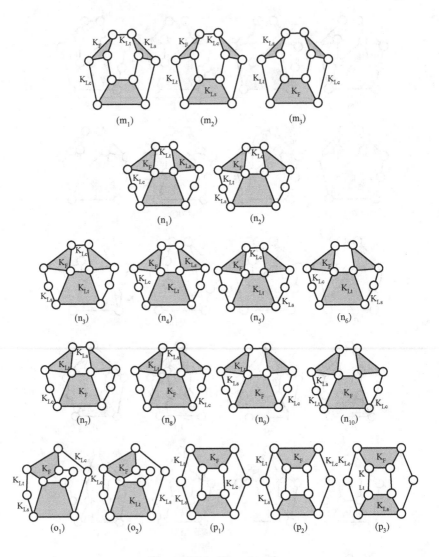

Figure 8.16 (*Continued*)

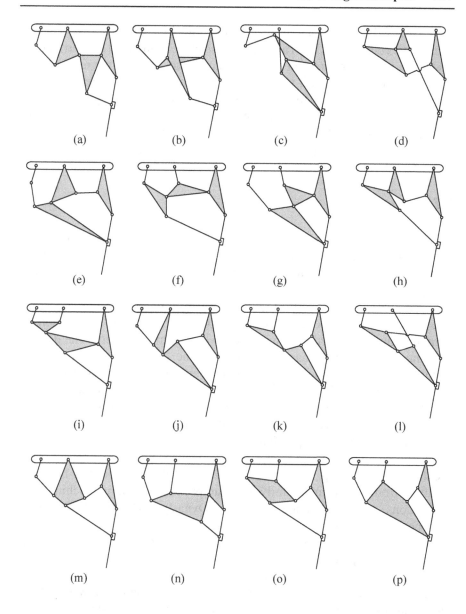

Figure 8.17   Some leg mechanisms for eight-bar type wooden horse carriages (Example 8.2)

Figure 8.18 shows a physical model developed at National Cheng Kung University (NCKU), Tainan, TAIWAN in 1996 [1]. This carriage is pushed to move forward and pulled to move backward. It requires only a small force to push or pull to walk up a reasonable slope. When left on a slope, it moves down without human intervention due to gravity. This might prove that such an invention may be feasible as indicated in the book Lun Heng by Wang Chong (AD ~27–97) in the Eastern Han Dynasty (AD 25–220), the earliest historical record of this legendary design.

8.7 Remarks

Lu Ban's wooden horse carriage is one of lost ancient Chinese machines with written descriptions but without illustrations and existing actual evidence. Restoration of this type is more difficult, and more imagination is required.

Evolutionists see history in terms of steady progress through the ages. The life cycle of a mechanical device is usually long, and its function is normally improved gradually with the development of society. The design of a mechanical device basically relied on the continuity of experience, and its development should be a continuous and smooth process. Therefore, if Lu Ban's wooden horse carriage in the Era of Spring and Autumn (770–481 BC) was real, Zhu-ge Liang's wooden ox and gliding horse invented in the Era of the Three Kingdoms (AD 220–280), 600–700 years later, could have been the eventual improvement of the original wooden horse carriage. Furthermore, 700–800 years later the mechanical horse for possible rice seedlings in the Song Dynasty (960–1279) could have been a heritage resulting from an artisan's family who had produced walking machines for generations. And, Jiang Yong of the Qing Dynasty (1644–1911) could also have been a member of the artisan's family.

Due to incomplete documentation and the loss of finished objects, the original structures of the walking machines, such as the wooden horse carriage of Lu Ban and the wooden ox and gliding horse of Zhu-ge Liang, have been a mystery through many eras. However, before new literature and/or hardware evidence are found, the proposed systematic approach for the generation of all possible topological structures of mechanisms subject to design requirements and constraints as described in Chapter 4 and Section 8.5 should be a novel direction and useful tool for assisting the study of lost ancient machinery. Thus, this chapter has sought in this spirit to trace the historical development of ancient Chinese walking machines.

Figure 8.18 NCKU's eight-bar-type wooden horse carriage [1]

References

1. Chiu, C.P., On the design of a wave gait walking horse (in Chinese), Master Thesis, Department of Mechanical Engineering, National Cheng Kung University, Tainan, Taiwan, June 1996.
   邱正平，波浪型步態機器馬之設計，碩士論文，國立成功大學機械工程學系，台南，台灣，1996 年 06 月。
2. Hwang, K., On the design of an optimal 8-link type walking horse (in Chinese), Master Thesis, Department of Mechanical Engineering, National Cheng Kung University, Tainan, Taiwan, June 1997.
   黃凱，最佳八連桿型機器馬之研究，碩士論文，國立成功大學機械工程學系，台南，台灣，1997 年 06 月。
3. Chen, P.H., On the mechanism design of 4-link and 6-link types wooden horse carriages (in Chinese), Master Thesis, Department of Mechanical Engineering, National Cheng Kung University, Tainan, Taiwan, June 1998.
   陳柏宏，四連桿與六連桿型木車馬之機構設計，碩士論文，國立成功大學機械工程學系，台南，台灣，1998 年 06 月。
4. Shen, H.W., On the mechanism design of 8-link type walking horses (in Chinese), Master Thesis, Department of Mechanical Engineering, National Cheng Kung University, Tainan, Taiwan, June 1999.
   沈煥文，八連桿型機器馬之機構設計，碩士論文，國立成功大學機械工程學系，台南，台灣，1999 年 06 月。
5. Yan, H.S., "A systematic approach for the restoration of the wooden horse carriage of ancient China," Proceedings of the International Workshop on History of Machines and Mechanisms Science, Moscow, Russian, pp. 199–204, 16–20 May 2005.
6. Yan, H.S., "Historical trace and restoration of ancient Chinese walking machines," Journal of the Chinese Society of Mechanical Engineers, Taipei, Vol. 26, No. 2, pp. 133–137, May 2005.
7. Xiao Yu-han, Stories of Lu Ban, Man Ting Fang Publisher, Taipei, 1994.
   蕭玉寒著，魯班傳奇，滿庭芳出版社，台北，1994 年。
8. Mo Zi (in Chinese) by Mo Di (Zhou Dynasty), Shanghai Ancient Books Publishing House, Shanghai, 1989.
   《墨子》；墨翟[周朝]撰，上海古籍出版社，上海，1989 年。
9. Huai Nan Zi (in Chinese) by Liu An (Han Dynasty), Shanghai Ancient Books Publishing House, Shanghai, 1989.
   《淮南子》；劉安[漢朝]撰，上海古籍出版社，上海，1989 年。
10. Lun Heng (in Chinese) by Wang Chong (Han Dynasty), Hong Ye Books, Taipei, 1983.
    《論衡》；王充[漢朝]撰，宏業書局，台北，1983 年。
11. Edited by Museum of Jing Men City, Bamboo Scripts of Guo Doan Chu Tomb (in Chinese), Cultural Subjects Publisher, Beijing, 1998.
    荊門市博物館編著，郭店楚墓竹簡，文物出版社，北京，1988 年。

12. Han Zi Yu Ping (in Chinese) by Han Fei (Zhou Dynasty), commented by Ling Ying-chu (Ming Dynasty), Foundation for Printing Famous Zi Xue Literatures of China, Taipei, 1978.
《韓子迂評》；韓非[周朝]撰/凌瀛初[明朝]評，中國子學名著集成編印基金會，台北，1978 年。

13. History of the Three Kingdoms (in Chinese) by Chen Shou (Jin Dynasty), Taiwan Commercial Press, Taipei, 1968.
《三國志》；陳壽[晉朝]撰，台灣商務印書館，台北，1968 年。

14. Tian Gong Kai Wu (in Chinese) by Song Ying-xing (Ming Dynasty), Taiwan Commercial Press, Taipei, 1983.
《天工開物》；宋應星[明朝]撰，天工開物，台灣商務印書館，台北，1983 年。

15. Nan Qi Shu · Zu Chong-zhi Zhuan (in Chinese) by Xiao Zi-xian (Liang Dynasty), Taiwan Commercial Press, Taipei, 1988.
《南齊書·祖沖之傳》；蕭子顯[梁]撰，南齊書，台灣商務印書館，台北，1988 年。

16. Tai Ping Imperial Panorama (in Chinese), edited by Li Fang (Song Dynasty), Taiwan Commercial Press, Taipei, 1983.
《太平御覽》；李昉[宋朝]等奉敕撰，台灣商務印書館，台北，1983 年。

17. Zhu-ge Liang Collection · Instructions on Making Wooden Ox and Gliding Horse (in Chinese) by Zhu-ge Liang (Three Kingdoms), Ding Wen Publishing House, Taipei, 1979.
《諸葛亮集·作木牛流馬法》；諸葛亮[三國]撰，諸葛亮集，鼎文出版社，台北，1979 年。

18. History of the Three Kingdoms · Biography of Zhu-ge Liang (in Chinese) by Chen Shou (Jin Dynasty), Taiwan Commercial Press, Taipei, 1968.
《三國志·諸葛亮傳》；陳壽[晉朝]撰，台灣商務印書館，台北，1968 年。

19. Yuan He Prefecture and County Policies (in Chinese) by Li Ji-fu (Tang Dynasty), Taiwan Commercial Press, Taipei, 1983.
《元和郡縣志》；李吉甫[唐朝]撰，台灣商務印書館，台北，1983 年。

20. Supplementary Notes on the History of the Three Kingdoms (in Chinese) by Hang Shi-Jun (Qin Dynasty), Commercial Press, Shanghai, 1936.
《三國誌補注》；杭世駿[清朝]撰，商務印書館，上海，1936 年。

21. History of the Three Kingdoms · Other Tales of Pu Yuan (in Chinese) by Chen Shou (Jin Dynasty), Taiwan Commercial Press, Taipei, 1968.
《三國志·蒲元別傳》；陳壽[晉朝]撰，台灣商務印書館，台北，1968 年。

22. Collective Analysis on the History of the Three Kingdoms (in Chinese) by Lu Bi, Yi Wen Publishing House, Taipei, 1958.
《三國志集解》；陳壽[晉朝]原著/盧弼[民國]撰，藝文出版社，台北，1958 年。

23. History of the Three Kingdoms · Book of Shu (in Chinese) by Chen Shou (Jin Dynasty), Taiwan Commercial Press, Taipei, 1968.
《三國志‧蜀書》；陳壽[晉朝]撰，台灣商務印書館，台北，1968 年。

24. The History of the Later Han Dynasty · History of the Political Units of Han Dynasty (in Chinese) by Yuan Shan-son (Jin Dynasty), Yi Wen Publishing House, Taipei, 1972.
《後漢書‧漢郡國志》；袁山松[晉朝]撰，藝文出版社，台北，1972 年。

25. Wei's Records of the Spring and Autumn Period (in Chinese) by Sun Sheng (Eastern Jin Dynasty), Editorial Committee on the Collection of Chinese History, Library, Si Chuan University, Chengdu, 1993.
《魏氏春秋》；孫盛[東晉]撰，中國史集成編委會，四川大學圖書館，成都，1993 年。

26. Inscriptions in the Zhu-ge Temple (in Chinese) by Shang Chi (Song Dynasty), collected in Tang Yun Cui by Yao Xuan (Song Dynasty), World Books, Taipei, 1972.
《諸葛武侯廟碑銘》；尚馳[唐]撰，收錄於唐文粹，姚鉉[宋]撰，世界書局，台北，1972 年。

27. Zi Zhi Tong Jian (Comprehensive Mirror for Aid in Government) (in Chinese) by Si-ma Guang (Song Dynasty), Jiu Zhou Publishing House, Beijing, 1998.
《資治通鑑》；司馬光[宋]撰，九洲出版社，北京，1998 年。

28. Book of General References (in Chinese) by Du You (Tang Dynasty), Taiwan Commercial Press, Taipei, 1987.
《通典》；杜佑[唐]撰，通典，台灣商務印書館，台北，1987 年。

29. Romance of the Three Kingdoms (in Chinese) by Luo Guan-zhong (Yuan Dynasty), Ba Shu Books Publishing House, Chengdu, 1995.
《三國演義》；羅貫中[元]撰，巴蜀書社，成都，1995 年。

30. Lin, K.L., On the improved design of Wang-Qian's wooden ox and gliding horse (in Chinese), Master Thesis, Department of Mechanical Engineering, National Cheng Kung University, Tainan, Taiwan, June 1995.
林寬禮，王淪木牛流馬之改進設計，碩士論文，國立成功大學機械工程學系，台南，台灣，1995 年 06 月。

31. Yan, H.S. and Lin, K.L., "A design of ancient China's wooden ox and gliding horse," Proceedings of the 10th World Congress on the Theory of Machines and Mechanisms, pp. 57–62, Oulu, Finland, 20–24 June 1999.

32. Yan, H.S., Creative Design of Mechanical Devices, Springer, Singapore, 1998.

Symbols

A_L	link assortment
C_{pi}	degrees of constraint of i-type joint for a planar mechanism
C_{si}	degrees of constraint of i-type joint for a spatial mechanism
D_V	revolute joints vertical to ground
D_H	revolute joints horizontal to ground
F	degrees of freedom
F_p	degrees of freedom for a planar mechanism
F_s	degrees of freedom for a spatial mechanism
G_p	nonrevolute joints incident to two members with perpendicular axial direction
G_e	nonrevolute joints incident to two members with parallel axial direction and with external connection
G_i	nonrevolute joints incident to two members with parallel axial direction and with internal connection
J_A	cam joint
J_C	cylindrical joint
J_F	flat joint
J_G	gear joint
J_J	pin-in-slot joint
J_O	rolling joint
J_P	prismatic joint
J_R	revolute joint
J_S	spherical joint
J_T	upper stopping joint
J_W	wrapping joint
J_X	fixed joint
K_A	cam
K_{Af}	follower
K_B	belt
K_C	chain
K_F	frame, ground link
K_G	gear
K_H	screw
K_I	piston

K_K	sprocket
K_{Li}	kinematic link of type i-type
K_O	roller
K_P	slider
K_R	rope
K_S	spring
K_T	actuator
K_U	pulley
K_W	wheel
K_Y	cylinder
L_i	link with i incident joints
m	maximum number of joints incident to a link
M_T	topology matrix
N_L	number of links or members
N_{Li}	number of links with i incident joints
N_J	number of joints
N_{Ji}	number of i-type joints

Index

History of Mechanism and Machine Science

1. M. Ceccarelli (ed.): *Distinguished Figures in Mechanism and Machine Science*. Their Contributions and Legacies, Part I. 2007
   ISBN 978-1-4020-6365-7
2. F.C. Moon: *The Machines of Leonardo Da Vinci and Franz Reuleaux*. Kinematics of Machines from the Renaissance to the 20th Century. 2007
   ISBN 978-1-4020-5598-0
3. H.-S. Yan: *Reconstruction Designs of Lost Ancient Chinese Machinery.* 2007
   ISBN 978-1-4020-6459-3